KNOW TOMORROW'S NEWS!

Biorhythms, the sensational new para-science that helps prepare its many practitioners for high-flying or disastrous "critical days" ahead, have become the new American pastime.

In *Biocycles*, biorhythm pioneer Vincent Mallardi takes the next giant step, from explanation of biorhythms to actual application—here at last is an easy method for using biorhythms to anticipate tomorrow's, or even next year's, news!

You can astonish your friends and be the life of any party by using this extraordinary new technique. You'll never want to stop "bio-forecasting" once everyone starts asking you "what's coming down?"

BIOCYCLES

YOU CAN USE BIORHYTHMS TO FORECAST YOUR FUTURE AND THAT OF YOUR FAVORITE NEWSMAKERS

VINCENT MALLARDI

A DELL BOOK

To Jill

Published by
Dell Publishing Co., Inc.
1 Dag Hammarskjold Plaza
New York, New York 10017

The author wishes to especially thank the many good friends
who assisted in the research and documentation
of news items in this book.
They include Pat Torelli, Ann Maloney, Joan Branfitt, Gerald Torelli,
Judy Maloney, Marcella Torelli, Sandra Gray, Cindy Pierog,
David Allman, Lisa Albaugh and William Kerr.

Copyright © 1978 by Vincent Mallardi

All rights reserved. No part of this book may be reproduced
or transmitted in any form or by any means,
electronic or mechanical, including photocopying,
recording or by any information storage and retrieval system,
without the written permission of the Publisher,
except where permitted by law.

Dell ® TM 681510, Dell Publishing Co., Inc.

ISBN: 0-440-10543-9

Printed in the United States of America

First printing—December 1978

Contents

Preface	9
Introduction	11
Biorhythm Tables	16
I. *THE EVENT FORETOLD*	26
The Carter Trip (1978)	26
Critical Days	29
Stevie Wonder (1973)	30
Sheila Young (1976)	31
Some *Physical* Actions	33

Muhammad Ali — Kareem Abdul-Jabbar
Evel Knievel — Richard Tucker
Yul Brynner — Willis Reed
Woody Allen — Donn Starry
Jimmy Brown — John Heinz
Carl Stokes — Russell Means
Mario Andretti — Garo Yepremian
Y. A. Tittle — Richard Nixon
George Wallace — Roman Gabriel
Lyndon Johnson — Peggy Fleming

Other Events on Physical Critical Days	41
Tale of Two Criticals	41

George McGovern and Fidel Castro

CONTENTS

Announcements from the Affective Personality ... 43
- Dixy Lee Ray
- Tom Brokaw
- Robert Sarnoff
- Gerald Ford
- John Chancellor
- Floyd McKissick
- Gale Sayers
- Claudine Longet
- Sheila Scott
- Pearl Bailey
- Milton Shapp
- Jack Nicklaus

Announcements of the Third Kind: The Objective ... 47
- Paul Hornung
- Princess Grace
- Richard Nixon
- Judy Carne
- Henry Kissinger
- Pierre Trudeau
- Marlon Brando
- Jason Robards, Jr.
- Gene Hackman
- Frank Sinatra
- Midge Costanza
- Earl Butz
- Norman Mailer
- James Farmer
- Cloris Leachman
- Spiro Agnew

When Many Critical Days Fall in a Row: Jimmy Carter ... 55

Mixed and Matched Criticals: Who Wins? ... 56
- Lou Graham and Hubert Green
- Bjorn Borg and Jimmy Connors

The Mystery of Necrobiosis ... 58

II. *WHERE NEWS STARTS* ... 60

Anticipating Rather than Following ... 60
- Steve Carlton
- Terry Bradshaw
- Lynn Swann

When Plans Go Awry ... 63
- Vladimir Horowitz
- Nelson Rockefeller
- Thomas P. Stafford
- George Meany
- A. J. Foyt

CONTENTS

Hubert Green
Richard Babbitt

When Should the Interview Take Place?   67
Gerald Ford

Critics Beware!   68
Bette Midler

People of the Theater   69

Washington, the New Hollywood   71
Benjamin Bradlee   Alan T. Howe
Wayne Hays   Joe D. Waggonner, Jr.
Carl Albert   Edwin A. Walker
Ron Nessen   Hamilton Jordan

The Media As the Makers   79
Barbara Walters   Howard Cosell
Tom Snyder   Dan Rather and
Morley Safer   Richard Nixon
Lauren Hutton   William Buckley
Marshall McLuhan

Media Personalities   85

III. THE SELECTION OF THE NEWSMAKERS   87

The Multiple Critical Days   87

The Double "Trouble" of a Double
   Critical Day   88

The Emotional/Intellectual Combination   89
Nadia Comaneci   Richard Daley
Andrew Young   Richard Pryor
Joe Namath   Roosevelt Grier
Harrison Schmitt

The Physical/Emotional Combination   93
Daniel Schorr   Joseph Califano
Roosevelt Grier   Gary Player
Jane Fonda   Leon Spinks
Julius Boros   F. Lee Bailey

	The Physical/Intellectual Combination	97
	G. Harrold Carswell Gerald Ford	
	Julius Erving Griffin Bell	
	Edward M. Kennedy William H. Rehnquist	
	Len Dawson Elvis Presley	
	Jerry Brown	
IV	THE ODYSSEY OF TRIPLE CRITICAL DAYS	104
	Rudolf Nureyev Pedro Borbon	
	Betty Ford Howard M. Goodman	
	Upcoming Triple Critical Days	109
V.	THE BIORHYTHM PING-PONG OF DIPLOMACY	112
	The SALT Talks	
	People in Government Service—U.S.	118
	International Newsmakers	121
	Other Newsmakers	122
	Who Will Be the New Newsmakers of the 80's?	124
Other Readings on Biorhythms and Performance Correlates		126

Preface

This is a book that will be immediately satisfying to you. It is a fun book to read and use, yet it is based on a very important scientific revelation: that each of us operates on a series of biological clocks that influence what we do from birth to death.

And with this book, biorhythms finally come of age.

Why? Because this is the first standardized way to compare any person's past actions with his/her likelihood for doing certain things in the future.

I've spent four years interviewing celebrities and everyday people. In the course of numerous television and lecture appearances, I have developed an easy technique for anyone to analyze past events and interpret likely future events.

By using this simple technique, you will be able to look up a person's permanent biorhythm values (based on date of birth) and then the date for which you want to know his or her biorhythms. It will then be possible to interpret that person's actions on performance for the given date. Or you may compare someone else's to yours, now or for the future!

Using this book, it's actually easy to determine how events are likely to go and thus feel the heady sensation of KNOWING TOMORROW'S NEWS.

You'll never want to stop applying this technique once people start asking you regularly, "What's coming down?"

Everybody has his ups and downs.

> . . . all aircraft accidents, attributable to either pilot factor or undetermined that have occurred within TAC since 1969 were analyzed (except for four for which no birth date of the pilots could be determined). The total sample was composed of 59 accidents wherein only the pilots involved were analyzed. Of those 59 accidents, 13 occurred on a critical biorhythmic day for at least one of the pilots involved.
>
> —TAC ATTACK.
> The Tactical Air Command, Langley AFB, VA.
> March 1972. 12:3.

> A major U.S. airline indicated that at present 25,000 of its 50,000 employees were being provided biorhythmic printouts. "This causes them to be a little more careful on a day when their curve bisects the halfway mark," an airline spokesman explained.
>
> —PLOTTING BIORHYTHMIC CURVES PAYS OFF.
> Article by Saunda Bentley
> *Philadelphia Inquirer*,
> August 24, 1975.

Introduction

Biorhythms have become the new American pastime: a kind of armchair tracking of important events through the use of a new tool of explanation.

Each day, when the news is headlined, what you are reading about are the actions of people who are influenced by the biorhythmic forces at work within them.

A diplomatic foul-up, for example, is reported in the *when* and *where* but usually the true reasons behind the episode are not explained. This is simply because the inner workings of the participants are not known!

This book is an attempt to correct this reporting defect by providing the reader with the material and techniques necessary to better explain the *whys* of current events, and some likely outcomes of future events.

The accuracy of such forecasts using biorhythms in this application is astounding. The outcome of sporting events may be predicted, sometimes even to the *exact* score. And it's really quite *un*mysterious. A historical record of the players' past performances is paired with the appropriate biorhythm time series so that correlations may be made. A qualitative assessment is completed, and the *tendency* (or probability) of the outcome is determined.

It is important to remember, however, that biorhythms do *not* guarantee outcomes, because they are internal to the participants. They *do* contribute to the tendency for victory or defeat in much the same way that new spark plugs and points contribute to the tendency for one race car to win over another, untuned machine. There are, of course, other factors, but at least one has been isolated.

As is common to all new prediction techniques, biorhythms are controversial. The uninformed deny that behavior is predictable, and the academic elite caution that they are years away from endocrinological mapping, a prerequisite for behavior tracking of internal origin. A third faction fears that verification of the technique will accelerate studies leading to personality modification.

The fact is that hormone treatments have been prescribed for decades in the control of diabetes and other functional disorders. More recently, female sex hormone extracts and synthetics have been used in birth control chemistries that "trick" the control mechanisms in the brain. The side effects of these hormonal interventions have included mood-modification and a case for a more thorough testing of hormone influences on human behavior.

While much is being written about the causes and associations of various biorhythms, this book is the logical next step, from causality to application.

Specifically, it deals with the time series commonly referred to as the combined biorhythms. The series includes a 23-day cycle of physical change that is based on desynchronizations between a person's internal *biological* clock and the *social* (24-hour regimen) clock. The common manifestation of losing or gaining an hour a day on the wall clock is a kind of jet-lag without the jet every 11–12 days, and an absence of stress every 23 days.

The emotional cycle covers 28 days and is based on periodic elevations of female sex hormones in both sexes and their interaction with external heat and light cues. The intellectual cycle of 33 days is

BIOCYCLES 13

based on similar periodic elevations of male sex hormones in both women and men.

In my earlier book (*Biorhythms & Your Behavior*, 1975), I brought together all of the theoretical and pragmatic literature on biological periodicity available at the time. That work, continually revised in each printing since then, has become the standard popular reference on biorhythms of *all* frequencies. I recommend it to all readers who are interested in formal explanations of causality.

In this new volume, I have changed notations with respect to "day references" so that all computations conform exactly to standard computer and electronic calculator programs.

Where formerly I referred to "a critical day" as "0" it is now "1." This substitution of integers does not change the accuracy of calculations in any of my books; the lapsed time remains the same.

I am indebted to the Running Press and the Synectics Network, both of Philadelphia, for the use of the following tables which are essential to the calculation of biorhythms.

By way of explanation, the reader should study the series of calculations that follow, and then calculate his or her own bio-values before going to the next chapter.

		Ph	Em	In
Vincent Mallardi				
born:	9/5	3	6	19
	42	13	6	24
		16	12	43

The biorhythm values for my birth date and year are on the following center spread. The date—September 5—is read 3 (Ph) 6 (Em) 19 (In). The year 1942 is

       Ph   Em   In
read 13   6   24. When the two sets are added, the totals must be reduced by as many 23's, 28's, and 33's as possible in the respective columns. Where the respective totals are *less than* 23, 28, or 33 no subtractions are necessary.

My biorhythm values, then, are 16 – 12 – 10.

For me, and any other person born on September 5, 1942, the *permanent* bio-values are 16 – 12 – 10. These figures may be added to any day in the future or past to determine my actual biorhythms at that precise time.

Let's try it for the second of July 1985.

	Ph	Em	In
My permanent net bio-values are:	16	12	10
July 1, 1985 (as computed from the date values tables, now add "1" to each value. This is for the following day, July 2.):*	9	6	23
	2	18	33

Now, the reader:

_____ born: day____
Name
                       year____ __ __ __

(less) all 23's, 28's, and 33's: __ __ __

Net bio-values: __ __ __

Make note of your permanent bio-values. In this way, you'll never have to compute them again.

---

*All date values in the tables are for the first day of each month. "1" must be added for each day during that month (e.g., July 10, add 9 to each date value).

BIOCYCLES 15

When you want to check a date—past or future—all you have to do is look up the date value in the tables and add it to the permanent values.

Try it for July 2, 1985:

	Ph	Em	In
Your permanent net bio-values are:	—	—	—
Date values:	—	—	—
The reader's biorhythms for that date:	—	—	—

In my case, the date indicates I will be *high* physically (2), *low* emotionally (18), and *low* intellectually (33). Also note that one day *earlier*, when my biorhythms would be (1)–17–33, I would experience a *critical day* physically.

Much of this book is about *critical days*, which are described in the next chapter. Later, you'll want to determine critical days in your own life, and verify the intriguing associations with personal newsmaking.

For now, here's a practice run on permanent biovalues for a more famous newsmaker:

		Ph	Em	In
Jimmy Carter				
born:	10/1/24			
	10/1–	0	8	26
	24–	9	0	31
		9	8	57
(Subtract all 23's, 28's, 33's if respective totals exceed them):		(–)	(–)	(33)
Net bio-values:		9	8	24

BIO-VALUES FOR DAY OF BIRTH

Day	January Ph Em In	February Ph Em In	March Ph Em In	April Ph Em In
1	20 1 2	12 26 4	7 26 9	22 23 11
2	19 0 1	11 25 3	6 25 8	21 22 10
3	18 27 0	10 24 2	5 24 7	20 21 9
4	17 26 32	9 23 1	4 23 6	19 20 8
5	16 25 31	8 22 0	3 22 5	18 19 7
6	15 24 30	7 21 32	2 21 4	17 18 6
7	14 23 29	6 20 31	1 20 3	16 17 5
8	13 22 28	5 19 30	0 19 2	15 16 4
9	12 21 27	4 18 29	22 18 1	14 15 3
10	11 20 26	3 17 28	21 17 0	13 14 2
11	10 19 25	2 16 27	20 16 32	12 13 1
12	9 18 24	1 15 26	19 15 31	11 12 0
13	8 17 23	0 14 25	18 14 30	10 11 32
14	7 16 22	22 13 24	17 13 29	9 10 31
15	6 15 21	21 12 23	16 12 28	8 9 30
16	5 14 20	20 11 22	15 11 27	7 8 29
17	4 13 19	19 10 21	14 10 26	6 7 28
18	3 12 18	18 9 20	13 9 25	5 6 27
19	2 11 17	17 8 19	12 8 24	4 5 26
20	1 10 16	16 7 18	11 7 23	3 4 25
21	0 9 15	15 6 17	10 6 22	2 3 24
22	22 8 14	14 5 16	9 5 21	1 2 23
23	21 7 13	13 4 15	8 4 20	0 1 22
24	20 6 12	12 3 14	7 3 19	22 0 21
25	19 5 11	11 2 13	6 2 18	21 27 20
26	18 4 10	10 1 12	5 1 17	20 26 19
27	17 3 9	9 0 11	4 0 16	19 25 18
28	16 2 8	8 27 10	3 27 15	18 24 17
29	15 1 7		2 26 14	17 23 16
30	14 0 6		1 25 13	16 22 15
31	13 27 5		0 24 12	

	May			**June**			**July**			**August**		
Day	Ph	Em	In	Ph	Em	In	Ph	Em	In	Ph	Em	In
1	15	21	14	7	18	16	0	16	19	15	13	21
2	14	20	13	6	17	15	22	15	18	14	12	20
3	13	19	12	5	16	14	21	14	17	13	11	19
4	12	18	11	4	15	13	20	13	16	12	10	18
5	11	17	10	3	14	12	19	12	15	11	9	17
6	10	16	9	2	13	11	18	11	14	10	8	16
7	9	15	8	1	12	10	17	10	13	9	7	15
8	8	14	7	0	11	9	16	9	12	8	6	14
9	7	13	6	22	10	8	15	8	11	7	5	13
10	6	12	5	21	9	7	14	7	10	6	4	12
11	5	11	4	20	8	6	13	6	9	5	3	11
12	4	10	3	19	7	5	12	5	8	4	2	10
13	3	9	2	18	6	4	11	4	7	3	1	9
14	2	8	1	17	5	3	10	3	6	2	0	8
15	1	7	0	16	4	2	9	2	5	1	27	7
16	0	6	32	15	3	1	8	1	4	0	26	6
17	22	5	31	14	2	0	7	0	3	22	25	5
18	21	4	30	13	1	32	6	27	2	21	24	4
19	20	3	29	12	0	31	5	26	1	20	23	3
20	19	2	28	11	27	30	4	25	0	19	22	2
21	18	1	27	10	26	29	3	24	32	18	21	1
22	17	0	26	9	25	28	2	23	31	17	20	0
23	16	27	25	8	24	27	1	22	30	16	19	32
24	15	26	24	7	23	26	0	21	29	15	18	31
25	14	25	23	6	22	25	22	20	28	14	17	30
26	13	24	22	5	21	24	21	19	27	13	16	29
27	12	23	21	4	20	23	20	18	26	12	15	28
28	11	22	20	3	19	22	19	17	25	11	14	27
29	10	21	19	2	18	21	18	16	24	10	13	26
30	9	20	18	1	17	20	17	15	23	9	12	25
31	8	19	17				16	14	22	8	11	24

	September			October			November			December		
Day	Ph	Em	In	Ph	Em	In	Ph	Em	In	Ph	Em	In
1	7	10	23	0	8	26	15	5	28	8	3	31
2	6	9	22	22	7	25	14	4	27	7	2	30
3	5	8	21	21	6	24	13	3	26	6	1	29
4	4	7	20	20	5	23	12	2	25	5	0	28
5	3	6	19	19	4	22	11	1	24	4	27	27
6	2	5	18	18	3	21	10	0	23	3	26	26
7	1	4	17	17	2	20	9	27	22	2	25	25
8	0	3	16	16	1	19	8	26	21	1	24	24
9	22	2	15	15	0	18	7	25	20	0	23	23
10	21	1	14	14	27	17	6	24	19	22	22	22
11	20	0	13	13	26	16	5	23	18	21	21	21
12	19	27	12	12	25	15	4	22	17	20	20	20
13	18	26	11	11	24	14	3	21	16	19	19	19
14	17	25	10	10	23	13	2	20	15	18	18	18
15	16	24	9	9	22	12	1	19	14	17	17	17
16	15	23	8	8	21	11	0	18	13	16	16	16
17	14	22	7	7	20	10	22	17	12	15	15	15
18	13	21	6	6	19	9	21	16	11	14	14	14
19	12	20	5	5	18	8	20	15	10	13	13	13
20	11	19	4	4	17	7	19	14	9	12	12	12
21	10	18	3	3	16	6	18	13	8	11	11	11
22	9	17	2	2	15	5	17	12	7	10	10	10
23	8	16	1	1	14	4	16	11	6	9	9	9
24	7	15	0	0	13	3	15	10	5	8	8	8
25	6	14	32	22	12	2	14	9	4	7	7	7
26	5	13	31	21	11	1	13	8	3	6	6	6
27	4	12	30	20	10	0	12	7	2	5	5	5
28	3	11	29	19	9	32	11	6	1	4	4	4
29	2	10	28	18	8	31	10	5	0	3	3	3
30	1	9	27	17	7	30	9	4	32	2	2	2
31				16	6	29				1	1	1

BIO–VALUES FOR YEAR OF BIRTH

Year	Ph	Em	In	Year	Ph	Em	In
1900	12	2	19	1948	6	26	10
01	15	1	17	49	9	25	8
02	18	0	15	50	12	24	6
03	21	27	13	51	15	23	4
04	0	25	10	52	17	21	1
05	3	24	8	53	20	20	32
06	6	23	6	54	0	19	30
07	9	22	4	55	3	18	28
08	11	20	1	56	5	16	25
09	14	19	32	57	8	15	23
10	17	18	30	58	11	14	21
11	20	17	28	59	14	13	19
12	22	15	25	60	16	11	16
13	2	14	23	61	19	10	14
14	5	13	21	62	22	9	12
15	8	12	19	63	2	8	10
16	10	10	16	64	4	6	7
17	13	9	14	65	7	5	5
18	16	8	12	66	10	4	3
19	19	7	10	67	13	3	1
20	21	5	7	68	15	1	31
21	1	4	5	69	18	0	29
22	4	3	3	70	21	27	27
23	7	2	1	71	1	26	25
24	9	0	31	72	3	24	22
25	12	27	29	73	6	23	20
26	15	26	27	74	9	22	18
27	18	25	25	75	12	21	16
28	20	23	22	76	14	19	13
29	0	22	20	77	17	18	11
30	3	21	18	78	20	17	9
31	6	20	16	79	0	16	7
32	8	18	13	80	2	14	4
33	11	17	11	81	5	13	2
34	14	16	9	82	8	12	0
35	17	15	7	83	11	11	31
36	19	13	4	84	13	9	28
37	22	12	2	85	16	8	26
38	2	11	0	86	19	7	24
39	5	10	31	87	22	6	22
40	7	8	28	88	1	4	19
41	10	7	26	89	4	3	17
42	13	6	24	90	7	2	15
43	16	5	22	91	10	1	13
44	18	3	19	92	12	27	10
45	21	2	17	93	15	26	8
46	1	1	15	94	18	25	6
47	4	0	13	95	21	24	4

DATE-VALUES

Month	1940 Ph Em In	1941 Ph Em In	1942 Ph Em In	1943 Ph Em In	1944 Ph Em In
Jan	19 19 3	17 21 6	14 22 8	11 23 10	8 24 12
Feb	4 22 1	2 24 4	22 25 6	19 26 8	16 27 10
Mar	10 23 30	7 24 32	4 25 1	1 26 3	21 28 6
Apr	18 26 28	15 27 30	12 28 32	9 1 1	6 3 4
May	2 28 25	22 1 27	19 2 29	16 3 31	13 5 1
Jun	10 3 23	7 4 25	4 5 27	1 6 29	21 8 32
Jul	17 5 20	14 6 22	11 7 24	8 8 26	5 10 29
Aug	2 8 18	22 9 20	19 10 22	16 11 24	13 13 27
Sep	10 11 16	7 12 18	4 13 20	1 14 22	21 16 25
Oct	17 13 13	14 14 15	11 15 17	8 16 19	5 18 22
Nov	2 16 11	22 17 13	19 18 15	16 19 17	13 21 20
Dec	9 18 8	6 19 10	3 20 12	23 21 14	20 23 17

Month	1945 Ph Em In	1946 Ph Em In	1947 Ph Em In	1948 Ph Em In	1949 Ph Em In
Jan	6 26 15	3 27 17	23 28 19	20 1 21	18 3 24
Feb	14 1 13	11 2 15	8 3 17	5 4 19	3 6 22
Mar	19 1 8	16 2 10	13 3 12	10 5 15	8 6 17
Apr	4 4 6	1 5 8	21 6 10	18 8 13	16 9 15
May	11 6 3	8 7 5	5 8 7	2 10 10	23 11 12
Jun	19 9 1	16 10 3	13 11 5	10 13 8	8 14 10
Jul	3 11 31	23 12 33	20 13 2	17 15 5	15 16 7
Aug	11 14 29	8 15 31	5 16 33	2 18 3	23 19 5
Sep	19 17 27	16 18 29	13 19 31	10 21 1	8 22 3
Oct	3 19 24	23 20 26	20 21 28	17 23 31	15 24 33
Nov	11 22 22	8 23 24	5 24 26	2 26 29	23 27 31
Dec	18 24 19	15 25 21	12 26 23	9 28 26	7 1 28

	1950	**1951**	**1952**	**1953**	**1954**
Month	Ph Em In	Ph Em In	Ph Em In	Ph Em In	Ph Em In
Jan	15 4 26	12 5 28	9 6 30	7 8 33	4 9 2
Feb	23 7 24	20 8 26	17 9 28	15 11 31	12 12 33
Mar	5 7 19	2 8 21	23 10 23	20 11 26	17 12 28
Apr	13 10 17	10 11 19	8 13 21	5 14 24	2 15 26
May	20 12 14	17 13 16	15 15 18	12 16 21	9 17 24
Jun	5 15 12	2 16 14	23 18 16	20 19 19	17 20 21
Jul	12 17 9	9 18 11	7 20 13	4 21 16	1 22 19
Aug	20 20 7	17 21 9	15 23 11	12 24 14	9 25 17
Sep	5 23 5	2 24 7	23 26 9	20 27 12	17 28 14
Oct	12 25 2	9 26 4	7 28 6	4 1 10	1 2 11
Nov	20 28 33	17 1 2	15 3 4	12 4 7	9 5 9
Dec	4 2 30	1 3 32	22 5 2	19 6 4	16 7 6

	1955	**1956**	**1957**	**1958**	**1959**
Month	Ph Em In	Ph Em In	Ph Em In	Ph Em In	Ph Em In
Jan	1 10 4	21 11 6	19 13 9	16 14 11	13 15 13
Feb	9 13 2	6 14 4	4 16 7	1 17 9	21 18 11
Mar	14 13 30	12 15 32	9 16 2	6 17 4	3 18 6
Apr	22 16 28	20 18 30	17 19 33	14 20 2	11 21 4
May	6 18 25	4 20 27	1 21 30	21 22 32	18 23 1
Jun	14 22 23	12 23 25	9 24 28	6 25 30	3 26 32
Jul	21 23 20	19 25 23	17 26 25	1 27 27	10 28 29
Aug	6 26 18	4 1 21	2 1 23	22 2 25	18 3 27
Sep	14 1 16	12 4 19	9 4 21	6 5 23	3 6 25
Oct	21 3 13	19 5 16	16 6 18	13 7 20	10 8 22
Nov	6 6 11	4 8 14	1 9 16	21 10 18	18 11 20
Dec	13 8 8	11 10 11	8 11 13	5 12 15	2 13 17

	1960	**1961**	**1962**	**1963**	**1964**
Month	Ph Em In	Ph Em In	Ph Em In	Ph Em In	Ph Em In
Jan	10 16 15	8 18 18	5 19 20	2 20 22	22 21 24
Feb	18 19 13	16 21 16	13 22 18	10 23 20	7 24 22
Mar	1 20 9	21 21 11	18 22 13	15 23 15	12 25 18
Apr	9 23 7	6 24 9	3 25 11	23 26 13	20 28 16
May	16 25 4	13 26 6	10 27 8	7 28 10	5 2 13
Jun	1 28 2	21 1 4	18 2 6	15 3 8	13 5 11
Jul	8 2 32	5 3 1	2 4 3	22 5 5	20 7 8
Aug	16 5 30	13 6 32	10 7 1	7 8 3	5 10 6
Sep	1 8 27	21 9 29	18 10 31	15 11 33	13 13 3
Oct	8 10 24	5 11 26	2 12 28	22 13 30	20 15 33
Nov	16 13 22	13 14 24	10 15 26	7 16 28	5 18 31
Dec	23 15 20	20 16 22	17 17 24	14 18 26	12 20 29

	1965	**1966**	**1967**	**1968**	**1969**
Month	Ph Em In	Ph Em In	Ph Em In	Ph Em In	Ph Em In
Jan	20 23 27	17 24 29	14 25 31	11 26 33	9 28 3
Feb	5 26 25	2 27 27	22 28 29	19 1 31	17 3 1
Mar	10 26 20	7 27 22	4 28 24	2 2 27	22 3 29
Apr	18 1 18	15 2 20	12 3 22	10 5 25	7 6 27
May	2 3 15	22 4 17	19 5 19	17 7 22	14 8 24
Jun	10 6 13	7 7 15	4 8 17	2 10 20	22 11 22
Jul	17 8 10	14 9 12	11 10 14	9 12 17	6 13 19
Aug	2 11 8	22 12 10	19 13 12	17 15 15	14 16 17
Sep	10 14 5	7 14 7	4 15 10	2 18 13	22 19 15
Oct	17 16 2	14 16 5	11 17 7	9 20 10	6 21 12
Nov	2 19 33	22 20 3	19 21 5	17 23 8	14 24 10
Dec	9 21 31	6 22 33	3 23 2	1 25 5	21 26 7

	1970	**1971**	**1972**	**1973**	**1974**
Month	Ph Em In	Ph Em In	Ph Em In	Ph Em In	Ph Em In
Jan	6 1 5	3 2 7	23 3 9	21 5 12	18 6 14
Feb	14 4 3	11 5 5	8 6 7	6 8 10	3 9 12
Mar	19 4 31	16 5 33	14 7 2	11 8 5	8 9 7
Apr	4 7 29	1 8 31	22 10 33	19 11 3	16 12 5
May	11 9 26	8 10 28	6 12 30	3 13 33	23 14 3
Jun	19 12 24	16 13 26	14 15 28	11 16 31	8 17 33
Jul	4 14 21	23 15 23	21 17 25	18 18 28	15 19 31
Aug	12 17 19	8 18 21	6 20 23	3 21 26	23 22 29
Sep	20 20 17	16 21 19	14 23 21	11 24 24	8 25 26
Oct	4 22 14	23 23 16	21 25 18	18 26 22	15 27 23
Nov	11 25 12	8 26 14	6 28 16	3 1 19	23 2 21
Dec	18 27 9	15 28 11	13 2 14	10 3 16	7 4 18

	1975	**1976**	**1977**	**1978**	**1979**
Month	Ph Em In	Ph Em In	Ph Em In	Ph Em In	Ph Em In
Jan	15 7 16	12 8 18	10 10 21	7 11 23	4 12 25
Feb	23 10 14	20 11 16	18 13 19	15 14 21	12 15 23
Mar	5 10 9	3 12 11	23 13 14	20 14 16	17 15 18
Apr	13 13 7	11 15 9	8 16 12	5 17 14	2 18 16
May	20 15 4	18 17 6	15 18 9	12 19 11	9 20 13
Jun	5 19 2	3 20 4	23 21 7	20 22 9	17 23 11
Jul	12 20 32	10 22 2	8 23 4	5 24 6	1 25 8
Aug	20 23 30	18 25 33	16 26 2	13 27 4	9 28 6
Sep	5 26 28	3 1 31	23 1 33	20 2 2	17 3 4
Oct	12 28 25	10 2 28	7 3 30	4 4 32	1 5 1
Nov	20 3 23	18 5 26	15 6 28	12 7 30	9 8 32
Dec	4 5 20	2 7 23	22 8 25	19 9 27	16 10 29

	1980	**1981**	**1982**	**1983**	**1984**
Month	Ph Em In	Ph Em In	Ph Em In	Ph Em In	Ph Em In
Jan	1 13 27	22 15 30	19 16 32	16 17 1	13 18 3
Feb	9 16 25	7 18 28	4 19 30	1 20 32	21 21 1
Mar	15 17 21	12 18 23	9 19 25	6 20 27	4 22 30
Apr	23 20 19	20 21 21	17 22 23	14 23 25	12 25 28
May	7 22 16	4 23 18	1 24 20	21 25 22	19 27 25
Jun	15 25 14	12 26 16	9 27 18	6 28 20	4 2 23
Jul	22 27 11	19 28 13	16 1 15	13 2 17	11 4 20
Aug	7 2 9	4 3 11	1 4 13	21 5 15	19 7 18
Sep	15 5 6	12 6 9	9 7 10	6 8 13	4 10 16
Oct	22 7 3	19 8 6	16 9 7	13 10 10	11 12 13
Nov	7 10 1	4 11 4	1 12 5	21 13 8	19 15 11
Dec	14 12 32	11 13 1	8 14 3	5 15 5	4 17 8

	1985	**1986**	**1987**	**1988**	**1989**
Month	Ph Em In	Ph Em In	Ph Em In	Ph Em In	Ph Em In
Jan	11 20 6	8 21 8	5 22 10	2 23 12	23 25 15
Feb	19 23 4	16 24 6	13 25 8	10 26 10	8 28 13
Mar	1 23 32	21 24 1	18 25 3	16 27 6	13 28 8
Apr	9 26 30	6 27 32	3 28 1	1 2 4	21 3 6
May	16 28 27	13 1 29	10 2 31	8 4 1	5 5 3
Jun	1 3 25	21 4 27	18 5 29	16 7 32	13 8 1
Jul	8 5 22	5 6 24	2 7 26	23 9 29	20 10 31
Aug	16 8 20	13 9 22	10 10 24	8 12 27	5 13 29
Sep	1 11 18	21 12 20	17 13 22	16 15 25	13 16 27
Oct	8 13 15	4 13 17	1 14 19	23 17 22	20 18 24
Nov	16 16 13	14 17 15	10 18 17	8 20 20	5 21 22
Dec	23 18 10	20 19 12	17 20 14	15 22 17	12 23 19

Month	1990 Ph Em In	1991 Ph Em In	1992 Ph Em In	1993 Ph Em In	1994 Ph Em In
Jan	20 26 17	17 27 19	14 28 21	12 2 24	9 3 26
Feb	5 1 15	2 2 17	22 3 19	20 5 22	17 6 24
Mar	10 1 10	7 2 12	5 4 14	2 5 17	22 6 19
Apr	18 4 8	15 5 10	13 7 12	10 8 15	7 9 17
May	2 6 5	22 7 7	20 9 9	17 10 12	14 11 14
Jun	10 9 3	7 10 5	5 12 7	2 13 10	22 14 12
Jul	17 11 33	14 12 2	12 14 4	9 15 7	6 16 9
Aug	2 14 31	22 15 33	20 17 2	17 18 5	14 19 7
Sep	10 17 29	7 18 31	5 20 33	2 21 3	22 22 5
Oct	17 19 26	14 20 28	12 22 30	9 23 33	6 24 2
Nov	2 22 24	22 23 26	20 25 28	17 25 31	14 27 33
Dec	9 24 21	6 25 23	4 27 26	1 28 28	21 1 30

Month	1995 Ph Em In	1996 Ph Em In	1997 Ph Em In	1998 Ph Em In	1999 Ph Em In
Jan	6 4 28	3 5 30	1 7 33	21 8 2	18 9 4
Feb	14 7 26	11 8 28	9 10 31	6 11 33	3 12 2
Mar	19 7 21	17 9 23	14 10 26	11 11 28	8 12 30
Apr	4 10 19	2 12 21	22 13 24	19 14 26	16 15 28
May	11 12 16	9 14 18	6 15 21	3 16 23	23 17 25
Jun	19 16 14	17 17 16	14 18 19	11 19 21	8 20 23
Jul	3 17 11	1 19 14	22 20 16	19 21 18	15 22 20
Aug	11 20 9	9 23 12	7 23 14	4 24 16	23 25 18
Sep	19 23 7	17 26 10	14 26 12	11 27 14	8 28 16
Oct	3 25 4	1 27 7	21 28 9	18 1 11	15 2 13
Nov	11 28 2	9 2 5	6 3 7	3 4 9	23 5 11
Dec	18 2 32	16 4 2	13 5 4	10 6 6	7 7 8

I. The Event Foretold

THE CARTER TRIP (1978)

When you look over the article below, you'll quickly realize that the event hadn't taken place as of the time it was written!

This news item was a published prediction based on a biorhythm situation for a then-future period.

> AS AIR FORCE ONE lifts off for his year-end summitry, President Carter will himself be crashing inversely into a down period, with a physical "critical" on his scheduled Dec. 29 departure. Intellectually, he'll be flying relatively high right through his meeting with the Shah of Iran on the 31st and India's Prime Minister Desai on the 2nd. The danger day for Jimmy: the 4th, with emotional and physical lows and a mental critical—just when he's eyeball-to-eyeball with the French president. (Of course, Giscard d'Estaing will be mighty low physically and intellectually himself that afternoon.)
>
> —from *People Weekly*, December 26, 1977.

BIOCYCLES 27

The biorhythm situation was easy to determine:

Jimmy Carter

	Ph	Em	In
born: 10/1/24			
bio-values (as previously worked out from tables):	9	8	24
date values (from date table, for Jan. 4, 1978):	10	14	26
	19	22	50

In this chart, the "Ph" stands for Physical, the "Em" for Emotional, and the "In" for Intellectual. When the birth date is looked up at the back of this book, the *bio-values* are found. (This applies to anyone who's alive!) Then, when a news date is looked up (on the date chart) and added to it, the biorhythm values are determined. Remember: when the total in any of the three series adds up to *more* than the number of days in the frequency, it's necessary to subtract all the multiples.

In the above example,

$$19 - 22 - 50$$

19 and 22 are each smaller totals than the maximum number of days (23 and 28) in their respective cycles. The Intellectual cycle total of 50, however, is *greater than* its maximum number of days (33). Therefore it is necessary to subtract 33 and arrive at

$$19 - 22 - ⑰$$

Graphically, President Carter's biorhythms looked like a downward zooming plane, crossing a Physical critical day in Poland, and an Intellectual critical day in France.

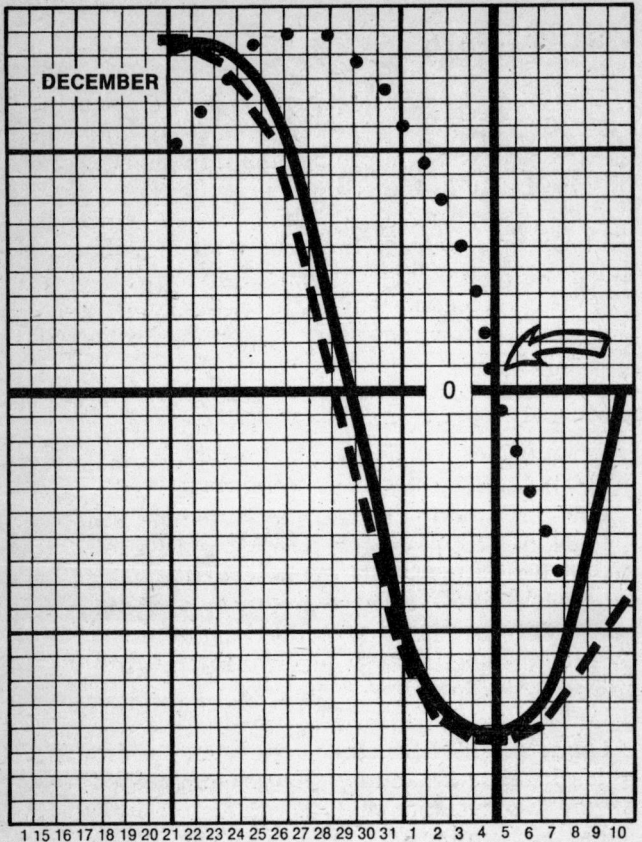

PRESIDENT CARTER'S BIORHYTHM GRAPH– DECEMBER 21, 1977, TO JANUARY 7, 1978*

The designation "critical day" refers to any point where the slope of the rhythm changes. This is at its point of inflexion with the "0" line. It is a period characterized by erratic and unusual performance, either very good or very poor.

* In biorhythm terminology, the following standard notations and symbols are used:

A Physical Cycle value:	Ph
An Emotional Cycle value:	Em
An Intellectual Cycle value:	In
A Physical Cycle illustrated:	solid line

CRITICAL DAYS

Critical days seem to coincide with newsmaking. The article in *People Weekly* essentially foretold what to expect, certainly not in detail but in general terms of *very good* and *very poor*.

With many months having passed since the Carter trip, it is interesting that news commentators still refer to the translation foul-up in Poland (on Physical critical date 12/29/77) and the "open microphone" embarrassment in India (on Intellectual critical date 1/3/78).

> **Steven Seymour,** the interpreter who misinterpreted some of **President Carter's** remarks on his December trip to Poland, remains actively employed by the State Department. He is currently translating Russian in Soviet-American talks in Geneva, department spokesmen confirmed yesterday. Seymour, a freelance interpreter, reportedly earns $150 a day for his services.

Carter's visit to France, on the other hand, was of little consequence and is scarcely remembered now. Maybe the President checked out his biorhythms in Paris, not wanting to repeat the previous days' embarrassments.

In serious terms, critical days are related to the concept of *stress*. As analyzed by the Canadian scientist and writer, Hans Selye, M.D., human beings possess a *finite* quantity of general adaptive energy.

An Emotional Cycle illustrated:	dashed line
An Intellectual Cycle illustrated:	dotted line
A critical day:	a number circled
Reference to a specific event day on a graph:	vertical line– horizontal line

Selye uses the abbreviation "G.A.S." (General Adaptive Syndrome) to describe this.

Like a reservoir, there is a given amount of general adaptive energy that, when used up, signals the end of life. The original amount of general adaptive energy is determined hereditarily.

The *use* of G.A.S. is described in a cycle consisting of
—alarm,
—response, and
—relaxation.

The points of *alarm* and *change* that take place between response and relaxation correspond to the critical day. In effect, the human being's behavior represents overreaction.

This extremity of action may be unconscious (physical), conscious (intellectual), or a combination of internal/external influences (emotional).

By plotting critical days the extremity in a given area of alarm may be predetermined.

Does it work? A few news items from the past illustrate the relationships.

STEVIE WONDER (1973)

SINGER STEVIE WONDER GOES INTO COMA

Salisbury, NC, Aug. 7, 1973—Rock singer Stevie Wonder is in a coma today, the result of an auto accident in which he was a passenger. Logs released from a truck rammed from behind by the singer's car, crashed through the windshield and struck Wonder on the forehead.

Stevie Wonder			
born: 5/13/50	*Ph*	*Em*	*In*
bio-values:*	15	5	8
date values: 8/7/73	9	27	32
biorhythms:	①	4	7

SHEILA YOUNG (1976)

> "(She's) the world's fastest female . . ."
>
> —Peter Schotting, quoted in *Sports Illustrated*, February 2, 1976

This statement by Sheila Young's coach was true when it was published on the eve of the 1976 Winter Olympics. The U.S. speed-skater had recently broken the world's record for 500 meters by going the distance in 40.91 seconds.

Something was "off," however, when she competed in her first Olympic race on Thursday, February 5, 1976:

---

*Remember the bio-values are found in the charts as follows:

```
5/13  -    3 -  9 - 2
1950  -   12 - 24 - 6
          15 - 33 - 8
             (28)
          15 -  5 - 8
```

SHEILA YOUNG PLACES 2ND . . .

Feb. 5, 1976—In her first Winter Olympic performance, world's record speed-skater Sheila Young failed to place first . . .

Something was "on" the following day:

YOUNG ESTABLISHES NEW OLYMPIC RECORD

Feb. 6, 1976—Sheila Young reversed her relatively poor performance yesterday by placing first in the 500-meter speed-skating race and establishing a new Olympic record at 42.76 seconds . . .

The biorhythm situation:

Sheila Young born: 10/14/50		Ph	Em	In	Ph	Em	In
bio-values:		22	19	19	22	19	19
date values:	2/5/76	1	15	20			
	2/6/76				2	16	21
		23	34	39	24	35	40
(less) multiples:		23	6	6	(1)	7	7

SOME PHYSICAL ACTIONS

In the following news summaries, only the biorhythms are shown. The bio-values and the date values that, added together, determine the actual biorhythms, are not shown. If the reader wishes to backtrack, the contributing bio-values may be found in the bio-values tables, pages 16–19. The contributing date values may be found in the date values tables, pages 20–25.

ALI ARRIVES IN LONDON AFTER HIS DEFEAT; ANNOUNCES A "NEW BATTLE PLAN."	Ph	Em	In
born: 1/18/42			
date: 2/17/78			
biorhythms:	①	20	13

EVEL KNIEVEL FAILS TO ROCKET 1600 FT. ACROSS SNAKE RIVER CANYON; SUFFERS CUTS AND SCRAPES; IS PULLED FROM CRASH.	Ph	Em	In
born: 10/17/38			
date: 9/8/74			
biorhythms:	①	7	10

YUL BRYNNER OPENS AGAIN IN NEW YORK: RAVE REVIEWS FOR *THE KING AND I.*

	Ph	Em	In
born: 7/11/20			
date: 10/7/77			
biorhythms:	①	20	19

COMEDIAN WOODY ALLEN PAYS FOR HIS LAUGHS; 500 STUDENTS AT $10 EACH DURING NEW FILMING.

	Ph	Em	In
born: 12/1/35			
date: 8/5/76			
biorhythms:	①	19	9

JIMMY BROWN TIES HIS SINGLE-GAME RUSHING RECORD OF 237 YARDS.

	Ph	Em	In
born: 2/17/36			
date: 11/19/61			
biorhythms:	①	28	3

BIOCYCLES 35

CARL STOKES DECLARES CANDIDACY FOR MAYOR OF CLEVELAND.

	Ph	*Em*	*In*
born: 6/21/27			
date: 6/16/67			
biorhythms:	①	18	20

MARIO ANDRETTI WINS INTERNATIONAL RACE OF CHAMPIONS AT DAYTONA; PLACES SECOND OVERALL IN 4-RACE SERIES.

	Ph	*Em*	*In*
born: 2/8/40			
date: 2/17/78			
biorhythms:	①	10	10

JIMMY BROWN SCORES 21 POINTS FOR SYRACUSE; TEAM GOES DOWN 28–27 TO TEXAS CHRISTIAN IN THE COTTON BOWL.

	Ph	*Em*	*In*
born: 2/17/36			
date: 1/1/57			
biorhythms:	⑫	9	2

Y. A. TITTLE USED "ALLEY-OOP" PASSES TO TALL R.C. OWENS IN SPECTACULAR WIN OVER BEARS.

	Ph	Em	In
born: 10/24/26			
date: 10/28/57			
biorhythms:	⑫	16	9

GOV. GEORGE WALLACE PHYSICALLY BARS PATH OF TWO NEGRO STUDENTS AS INTEGRATION IS ATTEMPTED.

	Ph	Em	In
born: 8/25/19			
date: 6/11/63			
biorhythms:	⑫	9	25

PRESIDENT JOHNSON PAYS QUICK VISIT TO PARATROOPERS; CRIES AT THEIR DEPARTURE TO WAR-TORN VIETNAM.

	Ph	Em	In
born: 8/27/08			
date: 2/17/68			
biorhythms:	⑫	24	10

BIOCYCLES 37

Eight years after this widely publicized event took place, author August Schara (*All the Presidents Plus*, 1978) wrote that LBJ had been duped: The planeload of troops was *not* bound for Vietnam but for another base in the United States.

> The commanders of the 82nd Airborne Division were notified only a couple of hours before the arrival of the President. Brg. Gen. Donald Blackburn, assistant commander of the division, was a key person in the cover-up, and later in 1975 finally told what really happened to Col. Hugh Robinson, a former military aide to LBJ. Later, in 1976, the (*Armed Forces*) *Journal* recanted the story of the Army's hoaxing.

Other examples of physically confused people continue, again shown with biorhythms totalled.

KAREEM ABDUL-JABBAR PUNCHES OPPOSING PLAYER; BASKETBALL SUPERSTAR FINED $5,000.	Ph	Em	In
born: 4/16/47			
date: 10/18/77			
biorhythms:	(12)	28	23

| | Ph | Em | In |

SINGER RICHARD TUCKER SUFFERS HEART ATTACK AFTER KALAMAZOO CONCERT.

born: 8/28/13

date: 1/8/75

biorhythms: ⑫ 14 7

BASKETBALL'S WILLIS REED BENCHED FOR REMAINDER OF SEASON DUE TO TENDONITIS.

born: 6/25/42

date: 11/10/71

biorhythms: ⑬ 7 6

GEN. DONN STARRY RECALLED FOR "SINO-SOVIET WAR" STATEMENTS; EXPLAINED AS "LAPSE OF JUDGMENT."

born: 5/31/25

date: 6/17/77

biorhythms: ⑬ 27 3

BIOCYCLES 39

SENATE CANDIDATE JOHN HEINZ SAYS DURING A DEBATE: A *SATURDAY NIGHT SPECIAL* **IS A PERSON WHO GOES OUT ON A SATURDAY NIGHT TO PURCHASE A HANDGUN!**

	Ph	Em	In
born: 10/23/38			
date: 9/8/76			
biorhythms:	(13)	4	9

AMERICAN INDIAN LEADER RUSSELL MEANS SEIZES CONTROL OF THE *MAYFLOWER.*

	Ph	Em	In
born: 11/10/39			
date: 11/26/70			
biorhythms:	(1)	28	21

GARO YEPREMIAN SETS ALL-TIME RECORD FOR MOST FIELD GOALS IN ONE QUARTER.

	Ph	Em	In
born: 6/2/44			
date: 11/13/66			
biorhythms:	(12)	24	16

RICHARD NIXON POSTPONES 4-PART INTERVIEW WITH DAVID FROST UNTIL AFTER FALL ELECTIONS.

    born: 1/9/13

    date: 1/10/76

    biorhythms: Ph ⑫   Em 24   In 1

ROMAN GABRIEL TIES ALL-TIME RECORD FOR MOST FUMBLES RECOVERED IN A GAME (HIS OWN AND OPPONENTS).

    born: 8/5/40

    date: 10/12/69

    biorhythms: Ph ⑫   Em 21   In 2

PEGGY FLEMING WINS HER FIRST WOMEN'S NATIONAL SENIOR SKATING TITLE.

    born: 7/27/48

    date: 1/11/64

    biorhythms: Ph ⑫   Em 19   In 4

OTHER EVENTS ON PHYSICAL CRITICAL DAYS

Muhammad Ali (1/18/42) defeats Ken Norton in rematch in Inglewood, CA, 9/10/73.

Vice-President Richard M. Nixon (1/9/13) fought his "kitchen" debate with Nikita Khrushchev in Moscow, USSR, 7/23/59.

Baseball's Sandy Koufax (12/30/35) pitched his third "no hitter," 6/4/64.

President Gerald Ford (7/14/13) asked Congress for a ban on handguns; then told them he was opposed to Federal registration or licensing of handguns, 6/19/75.

TALE OF TWO CRITICALS

Yes, it can happen. The odds are 36-to-1 against it, but occasionally two people—each with a physical critical day—get together. The results are likely to be distressing.

The problem is that the two aren't really "with it" as a pair, whereas either might be individually.

The results, in sporting activities, for example, are near draws. In show business and in politics (if there are differences), the two criticals often result in incongruous actions which are usually humorous in spite of the intentions of the characters.

Consider this first scene: A former presidential candidate is riding in a jeep in 100° weather through dusty streets. At the wheel is on-again, off-again enemy Fidel Castro. They stop, get out, and embrace for the cameras: A good media event with an outspoken liberal "dove" and a possibly mellowing Communist.

Before the news conference gets under way in this re-creation, the biorhythm picture should be known.

George McGovern   Ph   Em   In
born: 7/19/22
date: 5/7/75
biorhythms: (12)   22   14

Fidel Castro
born: 8/13/27
date: 5/7/75
biorhythms: (1)   19   11

Two physical criticals!

McGOVERN AND CASTRO CALL FOR "IMPROVED TIES" IN CONFUSING NEWS CONFERENCE

Havana, May 7, 1975—With U.S. Senator George McGovern standing beside him, Cuban Premier Fidel Castro turned from smiles to frowns and back again as he praised the Senator, denied involvement with John F. Kennedy's assassination, accused the U.S. Central Intelligence Agency of trying to kill him, and then said "We wish friendship."

ANNOUNCEMENTS FROM THE AFFECTIVE PERSONALITY

Unlike the preceding news items where the people operated out of a physical stamina change which occurs once every 11–12 days, critical days of the affective, or emotional, kind take place once every 14 days. These announcements are usually a bit more dramatic, sometimes with charges and inferences made.

DIXY LEE RAY QUITS AS ASST. SECRETARY OF STATE; CHARGES KEY OFFICIALS HAD NOT CONSULTED HER.

	Ph	Em	In
born: 9/3/14			
date: 6/19/75			
biorhythms:	10	①	29

TOM BROKAW ACCEPTS *TODAY SHOW* **OFFER; REPORTEDLY WILL RECEIVE $400,000 A YEAR.**

	Ph	Em	In
born: 2/6/40			
date: 6/8/76			
biorhythms:	2	①	7

ROBERT SARNOFF DECIDES TO QUIT TOP POST AT RCA.

	Ph	Em	In
born: 7/2/18			
date: 11/4/75			
biorhythms:	15	ⓛ	23

PRESIDENT FORD CALLS SEIZURE OF THE *MAYAGUEZ* AN "ACT OF PIRACY"; DEMANDS IMMEDIATE RELEASE.

	Ph	Em	In
born: 7/14/13			
date: 5/12/75			
biorhythms:	20	⑮	11

JOHN CHANCELLOR: "I HAD MONEY AND FAME . . . SO NOW IT'S TIME TO TAKE CONTROL OF MY LIFE"; ANNOUNCES PLAN TO LEAVE NBC.

	Ph	Em	In
born: 7/14/27			
date: 12/8/77			
biorhythms:	11	⑮	30

BIOCYCLES

FLOYD McKISSICK CALLS WHITE HOUSE CIVIL RIGHTS CONFERENCE A "HOAX."

	Ph	Em	In
born: 3/9/22			
date: 6/2/66			
biorhythms:	11	①	20

GALE SAYERS MATCHES ALL-TIME FOOTBALL RECORD OF MOST TOUCHDOWNS IN A SINGLE GAME; SCORES 6.

	Ph	Em	In
born: 5/30/43			
date: 12/12/65			
biorhythms:	22	①	16

CLAUDINE LONGET KILLS HER LOVER, VLADIMIR "SPIDER" SABICH; CONTENDS HE WAS SHOT ACCIDENTALLY.

	Ph	Em	In
born: 1/19/42			
date: 4/8/76			
biorhythms:	23	①	15

SHEILA SCOTT ATTEMPTS LONGEST SOLO FLIGHT;* IS "ENCOURAGED" BY TRAFFIC CONTROLLERS.

	Ph	Em	In
born: 4/27/27			
date: 5/18/66			
biorhythms:	7	⑮	11

PEARL BAILEY'S CHANGE OF LIFE? FORMER H.S. DROPOUT ENROLLS AT GEORGETOWN UNIVERSITY AT AGE OF 59.

	Ph	Em	In
born: 3/29/18			
date: 1/27/78			
biorhythms:	5	⑮	9

---

*Miss Scott became the first British pilot, man or woman, to fly solo around the world when, on June 12, 1966, she returned to London after having flown 32,000 miles.

> **PENNSYLVANIA'S GOV. MILTON SHAPP DINES AT THE BELLEVUE-STRATFORD; HOPED TO SHOW THERE WAS NO "LEGIONNAIRE'S DISEASE" IN THE FOOD.**
>
> | | | Ph | Em | In |
> |---|---|---|---|---|
> | *born:* | 6/25/12 | | | |
> | date: | 8/24/76 | — | — | — |
> | biorhythms: | | 23 | ① | 7 |

> **NICKLAUS WINS INVERRARY CLASSIC BY ONE STROKE; WAS "DISCOURAGED" UNTIL FINAL DAY WHEN HE CLOSED WITH FIVE STRAIGHT BIRDIES.**
>
> | | | Ph | Em | In |
> |---|---|---|---|---|
> | *born:* | 1/21/40 | | | |
> | date: | 2/26/78 | — | — | — |
> | biorhythms: | | 2 | ① | 24 |

ANNOUNCEMENTS OF THE THIRD KIND: THE OBJECTIVE

The third kind of change is the objective, or intellectual, which occurs once every 17–18 days. Statements and decisions made during these periods are often earth-shattering, in spite of the Jimmy Carter example cited earlier. Again, the figures you see below are biorhythm totals.

HALFBACK PAUL HORNUNG'S STRATEGY GIVES HIM 33 POINTS IN GREEN BAY WIN OVER BALTIMORE.

	Ph	Em	In
born: 12/23/35			
date: 10/8/61			
biorhythms:	15	14	⑰

PRINCESS GRACE MAKES HER FIRST AMERICAN STAGE APPEARANCE IN 26 YEARS; DESCRIBES HER TOUR OF POETRY READINGS "MERELY AN EVOLUTION" OF HER ACTIVITIES, NOT A COMEBACK.

	Ph	Em	In
born: 11/12/29			
date: 2/26/78			
biorhythms:	21	27	⑰

BIOCYCLES 49

> **NIXON MAKES FIRST STATEMENT SINCE RESIGNATION; RESPONDS TO PARDON WITH: "ONE THING I CAN SEE CLEARLY NOW IS THAT I WAS WRONG . . ."**
>
> | | *Ph* | *Em* | *In* |
> |---|---|---|---|
> | born: 1/9/13 | | | |
> | date: 9/8/74 | | | |
> | biorhythms: | 6 | 11 | ⑰ |

> **JUDY CARNE DENIES DRUG CHARGES; COPS DON'T LISTEN AND CONFISCATE BAG OF POWDER IN HER HOME.**
>
> | | *Ph* | *Em* | *In* |
> |---|---|---|---|
> | born: 4/27/39 | | | |
> | date: 2/15/78 | | | |
> | biorhythms: | 7 | 7 | ⑱ |

Pardon the interruption of the train of thought here, but the aftermath of the Judy Carne story is amusing. It seems that the sheriff's deputies seized a bag of powdered laxatives from the actress's Beverly Hills home, thinking it contained amphetamines. The embarrassed district attorney's office, unable to charge Miss Carne with a felony, still pushed for an arraignment date as a face-saving measure. The bill in Municipal Court was on a misdemeanor charge of possession of less than an ounce of marijuana. Admittedly, this is not one of the most earth-shaking stories of all time, but it is one which at least points out the incompetence of the intellectually low sheriff's deputies!

Let's go on.

	Ph	Em	In
HENRY KISSINGER POSTPONES SALT TREATY CONFERENCE; STATES HE "REALIZES THAT A SETTLEMENT ISN'T POSSIBLE AT THIS TIME."			
born: 5/27/23			
date: 12/9/75			
biorhythms:	8	10	⑰

	Ph	Em	In
MR. TRUDEAU SHUFFLES CABINET IN WHAT HE DESCRIBES AS "FAST HARD ACTION IN THE ECONOMIC AREA" AT A TIME WHEN CANADA IS "LIVING BEYOND ITS MEANS."			
born: 10/8/19			
date: 9/27/75			
biorhythms:	20	4	⑰

BIOCYCLES 51

MARLON BRANDO ADMITTED TO HOSPITAL FOLLOWING HIS "RIGHT HAND LEAD" TO A PHOTOGRAPHER'S JAW; DICK CAVETT LOOKS ON.

	Ph	Em	In
born: 4/3/24			
date: 6/13/73			
biorhythms:	6	21	⑰

ACTOR JASON ROBARDS CRITICALLY INJURED IN CRASH OF HIS CAR

Santa Monica, CA, December 8, 1972—Veteran stage actor Jason Robards, Jr., lies close to death in a hospital here after an auto accident in which his car skidded into the side of Encino Canyon Road. The actor was alone in his car.

Jason Robards, Jr.

	Ph	Em	In
born: 7/26/22	2	22	30
date: 12/8/72	20	9	21
biorhythms:	22	3	⑱

GENE HACKMAN HOSPITALIZED AFTER SUFFERING INJURIES ON MOVIE SET

London, November 20, 1976—Actor Gene Hackman suffered back and leg injuries today in a fall he incurred while filming here.

Gene Hackman

		Ph	Em	In
born:	1/30/31	20	20	22
date:	11/20/76	14	24	12
biorhythms:		11	16	①

FRANK SINATRA SUES EARL WILSON; CLAIMS BOOK ABOUT HIM IS GOOD AND BAD

Los Angeles, May 6, 1976—Singer Frank Sinatra is bringing suit against writer Earl Wilson over an unauthorized biography about Sinatra.

Sinatra stated that the book "although complimentary, is false, fictionalized, boring, and uninteresting."

Frank Sinatra			
	Ph	*Em*	*In*
born: 12/12/15	5	4	6
date: 5/6/76	0	22	12
biorhythms:	5	26	(18)

MIDGE COSTANZA'S REMARKS AT INTERNATIONAL WOMEN'S YEAR MEETING BRING STINGING REBUTTAL FROM GLORIA STEINEM.

Midge Costanza			
	Ph	*Em*	*In*
born: 11/28/32	19	24	14
date: 9/21/77	20	21	20
biorhythms:	16	17	(1)

AGRICULTURE SECRETARY EARL BUTZ UNDER FIRE FOR VULGAR RACIAL REMARKS MADE TO JOHN DEAN IN PRESENCE OF SINGER PAT BOONE.

Earl Butz			
	Ph	*Em*	*In*
born: 7/3/09	12	5	16
date: 8/19/76	13	15	18
biorhythms	2	20	(1)

Here are three more eye-opening episodes, with biorhythm value totals computed.

WRITER NORMAN MAILER ARRAIGNED ON STABBING OF WIFE.

	Ph	Em	In
born: 1/31/23			
date: 11/22/60			
biorhythms:	11	7	⑰

JAMES FARMER LEADS FREEDOM RIDERS' MARCH; ATTACKED BY MOBS IN BIRMINGHAM.

	Ph	Em	In
born: 1/12/20			
date: 5/14/61			
biorhythms:	11	7	⑱

CLORIS LEACHMAN RECEIVES HER FIRST OSCAR: BEST SUPPORTING ACTRESS.

	Ph	Em	In
born: 4/30/26			
date: 4/9/72			
biorhythms:	15	10	⑱

BIOCYCLES 55

There are many quitters in the news and, since such events often correspond to intellectual critical days 17 and 18, they deserve special notice.

FORMER VP TERMINATES DEAL, BLAMES "EXPLOITATION"

Baltimore, 2/6/75—Former U.S. Vice-President Spiro Agnew has decided to terminate a $100,000-a-year contract "in good faith" as a result of a land developer's alleged "exploitation" of Agnew. The other party, however, stated "I haven't received a dime from him and he's taken $75,000 from me."

Spiro T. Agnew			
	Ph	Em	In
born: 11/9/18	23	5	32
date: 2/6/75	5	15	19
biorhythms:	5	20	(18)

WHEN MANY CRITICAL DAYS FALL IN A ROW: JIMMY CARTER

Consider a very unusual situation: In rapid-fire order, an individual experiences several critical days in a row.

This would be an extremely confusing period for the person, and certainly not to his or her advantage.

The chances of such a string of days occurring are very remote—it may only happen a few times in a lifetime.

During 1977, however, the President of the United States made one of the silliest speeches of his political career. Jimmy Carter was seeking support for his turnaround on a campaign promise to balance the U.S. budget, a seemingly impossible task for U.S. leaders.

He prepared a speech and delivered it to a group of Democratic party leaders, about as friendly an audience as a Democratic president could pick. His tone had to be somber and statesmanlike, his logic impeccable.

But he had just entered a string of critical days:

		Ph	Em	In
preparation time	10/30/77	22	12	⑰
	10/31/77	23	13	⑱
delivery of the speech	11/1/77	①	14	19
	11/2/77	2	⑮	20

The resulting statement included a promise that "one billion dollars will go to Greyhound Bus . . . plus a whole bunch of tidbits." He also announced a postponement of all tax-reform proposals, another turnaround.

Confused? His friendly audience certainly was, as were millions who saw him on television later that evening.

MIXED AND MATCHED CRITICALS: WHO WINS?

Single critical days often bring on outstanding results and, in the case of professional athletes, it seems that the competitor who has the *last* critical comes in *first*.

BIOCYCLES 57

Lou Graham was the front-runner in the 1977 U.S. Open Golf Championship. Then, on the final day, June 18, Hubert Green came on strong and took the title by one stroke.

		Lou Graham			*Hubert Green*		
born:		1/7/38			12/28/46		
		Ph	Em	In	Ph	Em	In
bio-values:		16	6	29	5	5	19
date values:	6/15/77	14	7	21	14	7	21
biorhythms:	6/15/77	7	13	(17)	19	12	7
	6/16/77	8	14	(18)	20	13	8
	6/17/77	9	(15)	19	21	14	9
	6/18/77	10	16	20	22	(15)	10
	6/19/77	11	17	21	23	16	11

Let's look at another example:

BORG BEATS CONNORS IN EXTREMELY CLOSE CONTEST AT BOCA RATON TENNIS TOURNEY.

Bjorn Borg

		Ph	Em	In
born:	6/6/56			
date:	1/22/78			
biorhythms:		(12)	5	14

Jimmy Connors			
	Ph	Em	In
born: 9/2/52			
date: 1/22/78			
biorhythms:	5	6	①

THE MYSTERY OF NECROBIOSIS

On the ominous side of the critical day phenomenon is *necrobiosis*, the debilitating effect of stress on living tissue.

There is no formal, biomedical indication that the destruction of human tissue is predictable. It cannot be proven because the separation of tissue from any human being (while it does permit observation) necessarily removes that tissue from the origins of the stress.

Informal observation, however, indicates that many more people *die* on their critical days than chance would indicate.*

Necrologies are the recordings of deaths themselves (obituaries). Here is a list of necrologies for major newsmakers of the late twentieth century.

---

*Several tests of statistical significance have been applied to populations of individuals including studies utilizing major newspaper obituary notices without respect to cause of death. An experimental design that could avoid sampling errors for such things as "victim" deaths has yet to be developed primarily because most necrologies (obituaries) do not report causes of deaths except in the cases of famous people or when there are spectacular events. The problem is further complicated by the routine use of revival procedures and life-sustaining equipment after clinical death takes place.

Name	Born	Died	Biorhythms		
			Ph	Em	In
Stewart Alsop	5/17/14	5/26/74	6	①	13
Edward P. Biggs	3/29/06	3/10/77	17	⑮	10
Benjamin Britten	11/22/13	12/4/76	①	8	23
Bob Considine	11/4/06	9/25/75	①	19	11
Bing Crosby	5/2/04	10/15/77	⑫	6	①
Arthur John Daley	7/31/04	1/3/74	⑬	19	15
Dorothy Fields	7/15/05	3/28/74	①	6	14
Peter Finch	9/28/16	1/4/77	⑬	16	13
Peter Goldmark	12/2/06	12/7/77	18	11	①
Betty Grable	12/18/16	7/2/73	20	⑮	26
Bucky Harris	11/8/96	11/8/77	8	18	⑱
Hubert Humphrey	5/27/11	1/13/78	5	7	⑱
Estes Kefauver	7/26/03	8/10/63	⑫	7	19
Goddard Lieberson	4/5/11	5/29/77	⑫	26	6
Ted Mack	2/12/04	7/12/76	23	18	⑰
Archbishop Makarios III	8/13/13	8/3/77	22	⑮	3
Jayne Mansfield	4/19/37	6/29/67	⑫	25	7
Larry Parks	12/13/14	4/13/75	3	①	26
Rod Serling	12/25/24	6/28/75	2	24	①
Wernher Von Braun	3/23/12	6/16/77	22	27	①
Philip Wrigley	12/5/94	4/12/77	19	8	⑰
William Zeckendorf	6/30/05	9/30/76	⑬	14	22

II. Where News Starts

ANTICIPATING RATHER THAN FOLLOWING

In this century society has experienced an information explosion unlike anything experienced in the previous *thousand* centuries!

But news, like weather, is of the most value when it is *anticipated* rather than followed. In the pursuit of *likely* news, biorhythms may be used as a kind of crystal ball. The influence of biological rhythms on performance has been demonstrated on numerous occasions.

BIORHYTHMS EXPERT SAYS STEVE CARLTON SHOULDN'T PITCH

New York, Oct. 24, 1977 (AP)—Author Vincent Mallardi made a direct appeal to Philadelphia Phillies baseball manager Danny Ozark not to start National League leading pitcher Steve Carlton in tonight's playoff opener in Los Angeles.

Mallardi, who wrote *Biorhythms & Your Behavior*, indicated that Carlton will be in a physical "critical day" and won't do well against the Dodgers. He had little to say about Tommy John, except that he too will go out early in the game.

When I was interviewed for the above story, I had also determined that the Phillies' other pitchers would not be at their best according to the schedule Danny Ozark had arranged. In the case of Carlton, the situation was:

Steve Carlton	Ph	Em	In
born: 12/22/44			
date: 10/24/77			
biorhythms:	⑫	11	16

There's not a newsroom on the planet that won't use a good prediction-type "scoop" about an upcoming event. In the Carlton wire-service release to newspapers throughout North America, a reference was made to the short number of innings the writer thought Carlton would go before he was relieved in the National League playoffs that year.

During the game, Joe Garagiola, the TV announcer for the series, complimented the Philadelphia pitcher on his performance and ridiculed the writer for the content of the prediction. "So much for that theory!" he said, with a laugh.

A short while later Carlton was relieved. There was no comment from Joe.

BEHAVIOR

THOSE BIORHYTHMS AND BLUES

. . . Gil Brandt, vice-president of the Dallas Cowboys, is also convinced that biorhythm "has a lot of validity."

—*TIME, 2/27/78*

The evidence is that game results may be anticipated relative to principal performers.

PITTSBURGH STEELERS TAKE SUPERBOWL X ON SPECTACULAR TOUCHDOWN PASS

Miami, January 18, 1976—The Steelers beat Dallas by a close 21–17 when quarterback Terry Bradshaw made a 64-yard pass to wide receiver Lynn Swann.

Terry Bradshaw

	Ph	Em	In
born: 9/2/48	12	7	32
date: 1/18/76	6	25	2
biorhythms:	18	4	(1)

Lynn Swann

	Ph	Em	In
born: 3/7/52	18	13	4
date: 1/18/76	6	25	2
biorhythms:	(1)	10	6

WHEN PLANS GO AWRY

> ## VLADIMIR HOROWITZ CANCELS CLEVELAND CONCERT APPEARANCE; CITES ILLNESS
>
> Cleveland, March 5, 1978—Today's scheduled concert by pianist Vladimir Horowitz has been cancelled with no new date set. Horowitz, whose manager cites illness as the reason for the cancellation, was originally scheduled to perform on October 29 of last year. At that time, the concert was postponed until today's date. The reason then: illness.

> **Vladimir Horowitz**
>
> | | | Ph | Em | In | | Ph | Em | In |
> |---|---|---|---|---|---|---|---|---|
> | born: | 10/1/04 | | | | | | | |
> | bio-values: | | 0 | 5 | 3 | | 0 | 5 | 3 |
> | date: | 3/5/78 | 12 | 3 | 25 | | 1 | 18 | 20 |
> | biorhythms: | | ⑫ | 8 | 28 | | ① | 23 | 23 |

Most major news events are people-produced and are therefore *planned.* Illness, of course, isn't.

During media appearances in Cleveland during the week before the Horowitz concert, the author mentioned the biorhythm situation without predicting the cancellation. The biorhythm item became news. Then the announcement came.

Very often, a check of this type around the scheduled appearance of a well-known personality provides a warning beacon for the otherwise *un*expected turn of events. Without knowing what is going to happen, the reader is at least prepared for *something*, and won't be disappointed.

Time and time again, critical days interfere with the best-laid plans of everyone, maybe even mice as well as men! Let's look at some unusual changes of events during recent years:

VICE-PRESIDENT NELSON ROCKEFELLER GIVES "THE FINGER" IN RESPONSE TO HECKLERS

Binghamton, NY, January 5, 1977—Dubious history was made today when outgoing Vice-President Nelson Rockefeller gleefully raised his middle finger in a well-known obscene gesture directed at hecklers in this upstate New York community.

Nelson Rockefeller

	Ph	Em	In
born: 7/8/08	4	11	13
date: 1/5/77	14	4	25
biorhythms:	18	(15)	5

BIOCYCLES   65

ASTRONAUT CARRIED AWAY AS SPACECRAFT CLOSES IN ON THE MOON; CRIES OUT A FOUR-LETTER WORD

Houston, May 21, 1969—As millions around the world listened to Thomas P. Stafford as he guided Apollo 10's lunar landing module toward the moon's surface, there was sudden shock and a too-late attempt by NASA Mission Control to jam the General. In his awe and excitement, the astronaut, at 8 miles above the surface, yelled "F---!"

Gen. Thomas P. Stafford			
	Ph	Em	In
born: 9/17/30	17	15	25
date: 5/21/69	11	0	11
biorhythms:	5	⑮	3

LABOR LEADER GEORGE MEANY SAYS HE'S NON-PARTISAN; ASKS CBS FOR FREE-TIME POLITICAL TV SHOW	Ph	Em	In
George Meany			
born: 8/16/94			
date: 2/25/76	—	—	—
biorhythms:	17	⑮	13

66 VINCENT MALLARDI

> **INDY AUTO-RACE WINNER A. J. FOYT MAKES OBSCENE GESTURE DURING POCONO 500 RACE TRYOUTS; BANK SPONSOR MAY DROP HIM.**

A. J. Foyt			
	Ph	*Em*	*In*
born: 1/16/35	22	1	27
date: 6/23/77	21	14	28
biorhythms:	20	⑮	22

Here are some biorhythm summaries, with totals shown, for emotionally critical events.

SURPRISE GOLF TOURNEY ENDING; HUBERT GREEN WINS U.S. OPEN ONE SHOT AHEAD OF LOU GRAHAM.	*Ph*	*Em*	*In*
Hubert Green			
born: 12/28/46			
date: 6/18/77			
biorhythms:	22	⑮	10

> **MISSOURI POLITICAL LEADER'S EXTORTION TRIAL ENDS; RICHARD BABBITT CONVICTED ON 11 COUNTS OF MAIL FRAUD, 3 COUNTS OF EXTORTION, AND 1 COUNT OF ATTEMPTED EXTORTION; HAS NO COMMENT.**
>
> | | Ph | Em | In |
> |---|---|---|---|
> | **Richard Babbitt** | | | |
> | *born:* 10/30/35 | | | |
> | date: 7/27/77 | | | |
> | biorhythms: | 21 | ⑮ | ① |

WHEN SHOULD THE INTERVIEW TAKE PLACE?

Any good news reporter or concerned citizen now realizes that it's not *what* is said or meant, it is *when*, biorhythmically, it's said or done.

Biorhythms do funny things for (and to) people. Here's an example of triple low days (when people are "off" by about 15 to 25 percent). Note there are no critical values, just consistently *low* values.

> **GERALD FORD, ABOUT H. R. HALDEMAN'S BOOK,** *THE ENDS OF POWER*, much of which is about former President Ford:
> "It's all pretty interesting, but I don't understand it."
> **Gerald Ford**
>
> | | | Ph | Em | In |
> |---|---|---|---|---|
> | born: | 7/14/13 | | | |
> | date: | 2/17/78 | | | |
> | biorhythms: | | 20 | 19 | 33 |

CRITICS BEWARE!

Biorhythms also indicate that *criticals* may be far more important than *critics* when it comes to determining outcomes of play openings and their run-lengths.

Long before singer-comedienne Bette Midler opened at the Copacabana in 1978, the author was interviewed by a national magazine. The article in short said:

> **TO ERR IS HUMAN,** and no fans are more forgiving than those of the Divine Miss M. . . . Bette's lows would probably make others airsick, and (on) opening night her Copa should runneth over.
>
> *People Weekly*, 12/26/77

> **BETTE MIDLER COMPLAINS** of heat and "close encounters of the worst kind" in her Copa opening.
>
> All news services, 1/12/78

Bette Midler		Ph	Em	In
born:	12/1/45	6	5	15
date:	1/12/78	18	22	1
		(1)	27	16

PEOPLE OF THE THEATER

		Ph	Em	In
Richard Adler	8/3/21	14	15	24
Edward Albee	3/12/28	6	10	20
Harold Arlen	2/15/05	1	8	31
Elizabeth Ashley	8/30/39	14	22	33
Leonard Bernstein	8/25/18	7	25	9
Marlon Brando	4/3/24	6	21	7
Yul Brynner	7/11/20	11	11	16
Gower Champion	6/22/20	7	2	2
Marge Champion	9/2/23	13	11	23
Ilka Chase	4/8/05	18	12	12
Paddy Chayefsky	1/29/33	20	14	28
Sarah Churchill	10/7/14	22	15	8
Claudette Colbert	9/13/05	21	22	19
Betty Comden	5/3/19	9	26	22
Jackie Cooper	9/15/22	20	27	12
Hume Cronyn	7/18/11	10	23	4

PEOPLE OF THE THEATER

		Ph	Em	In
Jean Dalrymple	9/2/10	23	27	19
James Daly	10/23/18	17	22	16
Howard Da Silva	5/4/09	3	9	10
Joan Diener	2/24/34	3	19	23
J.P. Donleavy	4/23/26	15	27	16
Stephen Douglass	9/27/21	5	16	2
Alfred Drake	10/7/14	22	15	8
Faye Emerson	7/8/17	6	18	26
Alvin Epstein	5/14/25	14	7	30
Charles Farrell	8/6/06	16	3	22
Joey Faye	7/12/10	13	25	2
Jose Ferrer	1/8/12	14	28	13
Betty Field	2/8/18	21	27	9
Dorothy Fields	7/15/05	12	18	13
Nina Foch	4/20/24	12	4	23
Bob Fosse	6/23/27	3	21	19
Eddie Foy, Jr.	2/4/05	21	4	17
Martin Gabel	6/19/12	21	15	23
Frank Gagliano	11/18/31	4	8	27
Will Geer	3/9/02	17	18	16
Graham Greene	10/2/04	22	4	2
Helen Hayes	10/10/00	3	1	3
Patricia Hines	3/17/30	15	1	9
Israel Horovitz	3/31/39	5	6	10
John Houseman	9/22/02	4	17	17
Kim Hunter	11/12/22	8	25	20
William Inge	5/3/13	15	5	2
Burl Ives	6/14/09	8	24	2
Lou Jacobi	12/28/13	6	28	27
James Earl Jones	1/17/31	10	5	2
Elia Kazan	9/7/09	15	22	26
Howard Keel	4/13/19	6	18	9
Ruby Keeler	8/25/09	5	8	29
Jean Kerr	7/10/23	11	9	11
Walter Kerr	7/8/13	18	23	2
Richard Kiley	3/31/22	4	27	15
Alan Jay Lerner	8/31/18	1	19	3
Joshua Logan	10/5/08	2	24	23

PEOPLE OF THE THEATER

		Ph	Em	In
Mary Martin	12/1/13	10	17	21
Johnny Mercer	11/18/09	12	7	10
Burgess Meredith	11/16/08	11	10	14
Ethel Merman	1/16/09	19	5	19
Bob Merrill	5/17/20	20	10	5
Arthur Miller	10/17/15	14	3	28
John Nettleton	2/5/29	8	16	20
Mike Nichols	11/6/31	16	20	6
Jerry Orbach	10/20/35	21	4	14
Maureen O'Sullivan	5/17/17	12	14	12
Richard Rodgers	6/28/02	21	19	4
Stephen Sondheim	3/22/30	12	26	6
Gwen Verdon	1/13/26	23	15	17

WASHINGTON, THE NEW HOLLYWOOD

The term was perhaps first expressed by ABC Vice-President Brandon Stoddard. And today even the casual television viewer cannot avoid the plethora of political TV and motion picture dramas based on events not so far removed in time—past or future.

The opportunity to construct tomorrow's scenarios from recent events is assisted by biorhythms. Examine the way in which people initiate, and react to, the power exchanges that make news.

WASHINGTON POST CHIEF RESPONDS TO N.Y. TIMES ACCUSATION.

Washington, D.C., Feb. 24, 1978—Benjamin Bradlee, known for his Pulitzer Prize in Public Service for exposing the Watergate cover-up, had no comment on his own cover-up which the *Times* called "a second-rate burglary about a third-rate burglary."
The reference was to the *Post*'s securing of an advance copy of H.R. Haldeman's *The Ends of Power* by clandestine means.
Says Bradlee: "I'm happier than a pig in ___!"

	Ph	Em	In
born: 8/26/21			
date: 2/24/78			
biorhythms:	6	①	12

CONGRESSMAN WAYNE HAYS ANNOUNCES HE WILL RUN FOR PRESIDENCY.

Washington, D.C., March 22, 1976

Wayne Hays

	Ph	Em	In
born: 5/13/11	23	26	30
date: 3/22/76	1	5	0
biorhythms:	①	3	30

BIOCYCLES 73

> **SECRETARY WHO "CAN'T TYPE" CLAIMS SHE KEPT HER JOB HAVING SEX WITH WAYNE HAYS.**
>
> Washington, D.C., May 22, 1976

If anyone had been following Congressman Wayne Hays's biorhythms during 1976–1978, the way-out story of Hays's reactions to the blonde bombshell could almost have been foretold.

The Ohio congressman at first denied the allegations of his former assistant, but as publicity spread and further revelations were published, it became too much for the 65-year-old politician.

> **HAYS TAKEN TO A HOSPITAL UNCONSCIOUS AFTER SUFFERING WHAT DOCTORS CALL "OVERREACTION" TO AN OVERDOSE OF SLEEPING PILLS.**
>
> | | | Ph | Em | In |
> |------------|---------|-----|----|----|
> | born: | 5/13/11 | 23 | 26 | 30 |
> | date: | 6/10/76 | 12 | 1 | 14 |
> | biorhythms:| | (12)| 27 | 11 |

The critical-day pattern has continued ever since. Look at the biorhythm totals.

> **HAYS RELEASE FROM HOSPITAL DELAYED BECAUSE OF "STOMACH UPSET" COMPLAINT.**
>
> | | Ph | Em | In |
> |---|---|---|---|
> | date: 6/16/76 | | | |
> | biorhythms: | 18 | 5 | ⑰ |

> **HAYS CLAIMS OVERDOSE WAS "ACCIDENTAL," NOT SUICIDE; HE TOOK TEN TIMES THE PRESCRIBED DOSAGE.**
>
> | | Ph | Em | In |
> |---|---|---|---|
> | date: 7/4/76 | | | |
> | biorhythms: | ⑬ | 23 | 2 |

> **HAYS COLLIDES WITH A TRUCK WHILE DRIVING HIS OWN VEHICLE.**
>
> | | Ph | Em | In |
> |---|---|---|---|
> | born: 5/13/11 | 23 | 26 | 30 |
> | date: 9/7/76 | 9 | 6 | 4 |
> | biorhythms: | 9 | 4 | ① |

Hays retired from politics and disappeared from the Washington scene until early 1978 when he decided to run for Congress.

BIOCYCLES 75

born:	5/13/11	23	26	30
date:	2/22/78	13	7	9
biorhythms:		⑬	5	6

Washington, D.C., has many crazy events from bloopers to major personality revelations. A few follow in biorhythm totals form.

CARL ALBERT FURIOUS AT NEWS CONFERENCE AFTER HOUSE DEFEATS BILL THAT WOULD HAVE SENT 24 CONGRESSMEN TO ENGLAND TO FETCH THE MAGNA CARTA!

		Ph	Em	In
born:	5/10/08			
date:	3/16/76			
biorhythms:		⑫	3	33

WHITE HOUSE PRESS SECRETARY RON NESSEN CITES A RISE IN THE CONSUMER PRICE INDEX; SAYS PRESIDENT'S ECONOMIC POLICY IS WORKING!

		Ph	Em	In
born:	5/25/34			
date:	4/21/76			
biorhythms:		⑬	20	29

CONGRESSMAN ALAN T. HOWE ARRESTED BY 2 SALT LAKE CITY POLICE DECOYS POSING AS PROSTITUTES IN CITY'S RED-LIGHT DISTRICT.	*Ph*	*Em*	*In*

    born: 9/6/27

    date: 6/13/76

    biorhythms: ⑫   6   27

D.C. POLICE NAB CONGRESSMAN AFTER HE ALLEGEDLY SOLICITED POLICEWOMAN "PROSTITUTE." CONGR. JOE D. WAGGONNER, JR., IS RELEASED.	*Ph*	*Em*	*In*

    born: 9/7/18

    date: 1/19/76

    biorhythms: ①   10   32

BIOCYCLES

GEN. EDWIN A. WALKER ARRESTED FOR "FONDLING" IN MEN'S ROOM. DALLAS PARK POLICEMAN MAKES ARREST.	*Ph*	*Em*	*In*
born: 11/10/09			
date: 6/23/77			
biorhythms:	22	①	12

WHITE HOUSE AIDE IN BAR-SPITTING INCIDENT: DID HE OR DIDN'T HE?

Washington, D.C., February 17, 1978—The White House has been plagued with questions of whether or not aide Hamilton Jordan should resign over reports of hanky-panky type incidents during the past month. Jordan himself has indicated, "If the gossip columnists don't get me, I'll be around."

The gossip columnists certainly *did* try to get the young D.C. jet-setter when the *Washington Post* alleged he spat his drink down the front of the blouse of a young woman at a Georgetown bar. An unidentified female patron claimed that she started a conversation with the then recently separated 33-year-old power-broker.

"I asked him if he was Ham Jordan." She then reported that he tried to pick her up and, when she refused, spat his drink across her chest.

The incident was supposed to have taken place on Friday, January 28, 1978. The White House acknowledged that Ham had been in the tavern with a male friend: "They had stopped for a drink after a party." The President's office *denied* that any altercation with a woman ever took place, and the bartender corroborated the story.

Who was telling the truth?

Hamilton Jordan			
	Ph	Em	In
born: 9/21/44	5	21	22
date: 2/17/78	8	2	4
biorhythms:	(13)	23	26

While Jordan was indeed having a critical day when the story broke, he *wasn't* when he entered the Georgetown singles bar.

born:	9/21/44	5	21	22
date:	1/28/78	11	10	17
biorhythms:		16	3	6

He was, in fact, high emotionally and intellectually. Would the young lady please step forward and give the press her birthday and year?

There was another story—also speculation—about the White House man-about-town.

During a party on December 19, 1977, he reportedly turned to the attractive wife of the Egyptian ambassador. Standing slightly taller than she, he tugged at the neckline of her dress, and remarked that he had always wanted to see the twin pyramids of the Nile!

How was he that evening?

	born:	9/21/44	5	21	22
	date:	12/19/77	17	26	10
	biorhythms:		22	19	32

No one who was at the party was willing to comment. Notwithstanding the ambassador's wife (who may have thought the remark a compliment), Ham Jordan is biorhythmically innocent. As his alleged behavior suggests, he was *low* in all areas.

THE MEDIA AS THE MAKERS

A strange phenomenon has been described with concern by Daniel Schorr and other journalists. It is that people have become so drenched in cool media* such as television that they feel protected and insulated from the world as a result of the presence of a news "reporter." The word "reporter" is in quotes because it has lost much of its meaning. Most TV news is a combination of pop behavioral psychology, file tape, film footage, and cosmetics. It is style and not information. It is news that's already been synthesized and analyzed for the viewer like a prepared TV dinner: Don't ask how it was made, just take someone else's word for its accuracy.

---

*"Cool media" was coined by Marshall McLuhan, the Canadian communications commando, who describes cool media as essentially noninformational but very stylized, as opposed to the opposite: hot media. There is very little hot media left, I'm afraid.

The style has been sold as substance, and it has been bought. The "You Are There" type evening news formats are a pitiful joke, as are the so-called news magazines which, unfortunately, have chosen the road of direct competition with TV's style. Try to find "hard" news in any weekly (e.g., dates, times, quotes in context, etc.).

This simply means that, rightly or wrongly, the news*casters* (a much more appropriate term) are becoming news*makers* in the muddled perceptions of the public. As such, biorhythms enter the video picture because biorhythms are the essence of style. Stories, therefore, about newscasters follow the patterns that modern TV program V.P.'s encourage: identification between TV receiver (person) and TV deliverer (person).

Enough said. Herewith the newscasters.

BARBARA WALTERS BECOMES FIRST $1,000,000-A-YEAR NEWSPERSON; SIGNS WITH ABC-TV.

New York, October 21, 1976—It happened. Barbara Walters has accepted a contract that makes her the highest paid newscaster in the world. In accepting the ABC offer, she disputed the "circus atmosphere" statement made by NBC officials who implied she had been playing both ends in her negotiations.

Barbara Walters

		Ph	Em	In
born:	9/25/31	12	6	15
date:	10/21/76	10	9	22
biorhythms:		22	⑮	4

Here are other media examples in biorhythm totals form.

TOM SNYDER DECLINES TO READ BARBARA WALTERS DIVORCE STORY OVER TV NEWS SHOW.

	Ph	Em	In

While Barbara Walters's picture flashed on the screen, Snyder ignored the copy and reported instead on the Princess Margaret–Anthony Snowden separation.

Tom Snyder
   born: 5/12/36

   date: 3/23/76

   biorhythms: 2   ①   8

MORLEY SAFER SAYS BARBARA WALTERS IS "NO REPORTER," THAT SHE STOPPED BEING ONE WHEN SHE TOLD JIMMY CARTER "BE WISE WITH US . . . BE GOOD TO US." WALTERS ANGRILY RESPONDS.

	Ph	Em	In

Morley Safer
   born: 11/8/31

   date: 12/20/76

   biorhythms: ⑫   16   13

Obviously Barbara Walters is news, and this display could go on forever, *ad nauseum*.

There are other newscasters, though, and their incidents at least as interesting.

IT WAS A VERY BAD DAY ON "GOOD DAY" MORNING TALK SHOW

Boston, April 13, 1977—Millions of listeners were shocked, and switchboards around New England lit up like Christmas trees when a guest on the #1 rated morning TV news and information show used obscene language. Because the program is carried "live" on the New England Network, studio censors could not flip the off-switch in time.

The author has been one of hundreds of media people that have been interviewed on the "Good Day" Show which is broadcast "live" around the country on a syndicated basis.

One of the hosts, Janet Langhart, was interviewing fashion-model, show business personality Lauren Hutton on April 13, 1977. It was one of those "how did you make it to the top" type questions that was being routinely asked; the kind of question that's supposed to normally elicit a response of a bragging talenty nature.

Langhart is a good journalist and, trying to avoid the trite, apparently asked the question in a way that intimated the stereotyped loss-of-virtue scenario.

Lauren Hutton's answer to how she got to the top:

"Actually, I f---ed around a lot!"

Lauren Hutton	Ph	Em	In
born: 11/17/43	15	22	1
date: 4/13/77	19	0	24
biorhythms:	⑫	22	25

HOWARD COSELL ATTACKED BY BRICKS IN TAVERN

Wildwood, N.J.—Cafe owner Charles Brand says that business has boomed since he started Cosell brick-throwing contests at his establishment. It seems that whenever Howard Cosell appears on the TV screen, one patron is allowed to throw a brick at the sportscaster's image.

Brand says they're all old TV sets.

Howard Cosell is an event created by the medium he uses to cover real events. Since he is likely to continue his pseudo-newsmaking binge, his vital biorhythm statistics follow. Maybe the reader would like to lay a brick on this page!

Howard Cosell	Ph	Em	In
born: 3/25/29	6	24	5
date values: ?	—	—	—
biorhythms:			

Dan Rather's biorhythm total indicated a critical physical day when he became involved in a highly publicized and long running altercation with then-President Richard Nixon.

> **DAN RATHER DENIES BAITING NIXON; SAYS "TIJUANA BULL RING" NEWS CONFERENCE MOOD WAS CREATED BY THE PRESIDENT, NOT HIM.**
>
> | | Ph | Em | In |
> |---|---|---|---|
> | *born:* 10/31/31 | | | |
> | date: 10/31/73 | | | |
> | biorhythms: | ① | 26 | 30 |

Incidentally, in support of Dan Rather, President Nixon *was* in an intellectual critical day at the news conference.

President Nixon

	Ph	Em	In
born: 1/9/13	14	7	17
date: 10/31/73	1	27	17
biorhythms:	15	6	①

Here's an "important" bit of newsmaking by a media person. The biorhythm totals illustrate an intellectual critical day.

		Ph	Em	In
PUBLISHER BILL BUCKLEY BREAKS COLLARBONE ON HIS YACHT.				
William Buckley				
born:	11/24/25			
date:	8/2/68			
biorhythms:		22	25	⑰

MEDIA PERSONALITIES

Permanent Bio-Values

		Ph	Em	In
Cleveland Amory	9/2/17	19	18	3
Jack Anderson	10/19/22	9	21	11
Roone Arledge	7/8/31	22	1	28
Joey Bishop	2/13/18	16	22	4
Frank Blair	5/30/15	17	4	4
David Brinkley	7/10/20	12	12	17
Tom Brokaw	2/26/40	14	1	27
Helen Gurley Brown	2/18/22	22	12	23
Anita Bryant	3/25/40	13	10	13
John Chancellor	7/14/27	20	25	0
Dick Clark	11/30/29	9	26	19
Howard Cosell	3/25/29	6	24	5
Walter Cronkite	11/4/16	22	12	8
Bill Cullen	3/18/20	12	14	32
Mike Douglas	8/11/25	17	2	7
Hugh Downs	2/14/21	0	17	29

MEDIA PERSONALITIES

Permanent Bio-Values

		Ph	Em	In
Michael Fox	8/13/37	2	13	11
David Frost	4/7/39	21	27	3
Betty Furness	1/3/16	5	9	16
Joe Garagiola	2/12/26	16	13	20
Jim Hartz	2/3/40	17	4	30
Hugh Hefner	4/9/26	6	13	30
Sonny Jurgensen	8/23/34	7	7	8
Bert Lance	6/3/31	11	8	30
John Lindsay	11/24/21	16	14	10
Art Linkletter	7/17/12	6	15	28
Frank McGee	9/12/21	20	3	17
Jim McKay	9/24/21	8	19	5
Ed McMahon	3/6/23	9	23	5
Roger Mudd	2/9/28	2	14	19
Keith Murdoch	3/11/31	3	8	15
William S. Paley	9/28/01	18	12	13
Dan Rather	10/31/31	22	26	12
Gene Rayburn	12/22/17	0	19	24
Harry Reasoner	4/17/23	13	9	29
James Reston	11/3/09	4	22	25
Morley Safer	11/8/31	14	18	4
Pierre Salinger	6/14/25	6	4	32
Phyllis Schlafly	8/15/24	10	27	5
Daniel Schorr	8/31/16	18	21	7
Tom Snyder	5/12/36	0	5	7
Carl Stokes	6/21/27	5	23	21
Forrest Tucker	2/12/19	20	22	3
Sander Vanocur	1/8/28	10	17	17
Mike Wallace	5/9/18	0	21	18
Barbara Walters	9/25/31	12	6	15

III. The Selection of the Newsmakers

The best way to select the people who "make news biorhythmically" is to determine in advance when various likely newsmakers are "scheduled" for *double critical* and *triple critical* days.

Because of the relative infrequency of these multiple critical days, their occurrences are pseudo-news events in themselves. Add to this the associations of historical events with other multiple criticals, and there emerge occasions for close watching of the person and the events in which he or she is playing a role.

THE MULTIPLE CRITICAL DAYS

Multiple critical days occur when more than one rhythm are either at the midpoint or beginning of their respective cycles.

Double critical days (when two are in these positions) occur about 10 times each year. The frequencies and combinations are:

	Ph	Em	In
Physical & Intellectual—	12		17
3 times/year	13		17
	12		18
	13		18

Physical & Intellectual
Combinations—
4 times/year

1	1
1	17
1	18
12	1
13	1

Emotional & Intellectual
Combinations—
3 times/year

1	1
14	1
1	17
1	18
14	17
14	18

Physical & Emotional
Combinations—
3 times/year

1	1
1	14
12	1
13	1
12	14
13	14

Triple critical days occur far less frequently. Because of their rarity, triple critical days are treated separately in Chapter IV.

THE DOUBLE "TROUBLE" OF A DOUBLE CRITICAL DAY

"Trouble" can be good or it can be bad. Unlike the relatively common *single* critical days, *double* critical days usually lead to much more dramatic departures from so-called "normal" behavior.

In some respects, the *double* critical day is for *one* person as dramatic as a *single* critical day is for *two* people working together (Chapter II).

Judgments about which combinations of double criticals are "good" and which are "bad" must be

BIOCYCLES 89

based upon the reader's experiences. In the array of news documentation available in this work, the combinations do fall into general categories of "good" and "bad." These are empirical observations, however, and are far from conclusive as yet.

THE EMOTIONAL/INTELLECTUAL COMBINATION

The simultaneous change of emotional and mental dispositions appears, tentatively, to be *positively* dramatic: a sort of combination mood and psyche job being done by the person on him or herself.

NADIA COMANECI DOES "THE IMPOSSIBLE"

Montreal, July 18, 1976—Nadia Comaneci, the 14-year-old Rumanian gymnast, did what most Olympic observers thought was "impossible." She received a perfect score of ten today.

Nadia Comaneci		Ph	Em	In	Ph	Em	In
born:	11/12/61	13	4	31	13	4	31
date values:	7/18/76	14	11	19			
	7/19/76	—	—	—	15	12	20
biorhythms:		4	(15)	(17)	5	16	(18)

Nadia repeated her perfect score performance the following day during an intellectual critical. Note her biorhythm totals.

U.N. AMBASSADOR ANDREW YOUNG SAYS CUBANS HAVE "STABILIZING" ROLE IN ANGOLA; "FREE WHEELING" REMARKS DRAW IRE FROM CRITICS.

	Ph	Em	In
born: 3/12/32			
date: 3/26/77			
biorhythms:	6	⑮	⑰

The controversial Andrew Young performed on a battery of *verbal* gymnastics in the days following his "Cuban" remarks. His comments included the famous "racist" allegations against most groups and societies.

A "MELANCHOLY" JOE NAMATH QUITS

Ft. Lauderdale, FL, Jan. 23, 1978—Joe Namath is, in his own words, "not going to play next year." The 34-year-old veteran quarterback for the Los Angeles Rams is best known for his 13 seasons with the N.Y. Jets. Namath told *N.Y. Times* reporter Dave Anderson: "It was no fun being a second string quarterback. Sometimes it was a bit melancholy . . ."

Joe Namath

	Ph	Em	In
born: 5/31/43	1	24	6
date: 1/23/78	6	5	12
biorhythms:	7	①	⑱

Here are some other double criticals in biorhythm totals.

	Ph	Em	In
ASTRONAUT HARRISON SCHMITT DISCOVERS COLORED SOIL ON MOON; IRON OXIDE POSSIBILITY SUGGESTS WATER IN MOON'S HISTORY.			
born: 7/3/35			
date: 12/13/72			
biorhythms:	17	⑮	⑰

	Ph	Em	In
CHICAGO MAYOR RICHARD DALEY ORDERS NATIONAL GUARD TO "SHOOT TO KILL" DEMONSTRATORS AT DEMOCRATIC CONVENTION. POLICE SHOOT HIPPIE TO DEATH.			
born: 5/15/02			
date: 8/22/68			
biorhythms:	11	⑮	⑱

RICHARD PRYOR ARRESTED IN SHOOTING INCIDENT.

Los Angeles, February 5, 1978—Comedian Richard Pryor was arrested and charged with shooting at two of his wife's friends. He allegedly also rammed their car with his Mercedes-Benz.

Richard Pryor			
	Ph	Em	In
born: 12/1/40	15	11	26
date: 2/5/78	19	18	25
biorhythms:	11	(1)	(18)

ROBERT KENNEDY'S ASSASSIN TACKLED BY FOOTBALL STAR ROOSEVELT GRIER; WALKING WITH ETHEL KENNEDY WHEN SHOTS RANG OUT, GRIER CAUGHT SUSPECT SIRHAN SIRHAN AND HELD HIM IN A BEAR-LOCK UNTIL POLICE CAME.

Roosevelt Grier			
	Ph	Em	In
born: 7/14/32	18	21	19
date: 6/6/68	14	22	32
biorhythms:	9	(15)	(18)

THE PHYSICAL/EMOTIONAL COMBINATION

When the body and the spirit change together there is a likelihood that news will be an unexpected and dramatic response to outside changes: a kind of spiritual kick of oneself into another gear.

One such example was the change of CBS reporter Daniel Schorr. He was covering investigations of the U.S. House Intelligence Committee on illegal CIA and FBI operations. On the weekend of January 24–25, 1976, Schorr broadcast news items on the basis of a photocopied report he had obtained of a secret session.

On Thursday, following a House vote of 246 to 124 against making any such report public, Schorr realized that he had the only copy of the suppressed document.

He decided to publish it.

Daniel Schorr

		Ph	Em	In
born:	8/31/16	18	21	7
date:	1/29/76	17	8	13
biorhythms:		(12)	(1)	20

Physical/Emotional biorhythm totals follow for a few more famous faces.

FOOTBALL'S ROOSEVELT GRIER ANNOUNCES SINGING DEBUT AT CARNEGIE HALL.

	Ph	Em	In
born: 7/14/32			
date: 1/17/63			
biorhythms:	(13)	(1)	24

JANE FONDA ANNOUNCES SHE WILL GO TO NORTH VIETNAM ON PEACE INITIATIVE.

	Ph	Em	In
born: 12/21/37			
date: 2/16/71			
biorhythms:	(13)	(15)	33

JULIUS BOROS WINS U.S. OPEN GOLF CHAMPIONSHIP.

	Ph	Em	In
born: 3/3/20			
date: 6/11/52			
biorhythms:	(13)	(1)	8

**HEW SECRETARY
JOSEPH CALIFANO
TOUCHES OFF
CONTROVERSY BY
LOOKING
"FAVORABLY" ON
FAVORABLE
TREATMENTS FOR
BLACKS IN
UNIVERSITY
ADMISSIONS.**

Ph   Em   In

    born: 5/15/31

    date: 3/18/77

    biorhythms: ①   ①   14

**GARY PLAYER WINS
THE MASTERS
TOURNEY IN A
SURPRISE UPSET OVER
HUBIE GREEN.**

Ph   Em   In

    born: 11/1/36

    date: 4/9/78

    biorhythms: ①   ⑮   21

Golf fans will recall that Player was behind Green and Watson until the final round when Player came up strong and Watson fell behind Green, whose biorhythms were 18–2–8, missed an easy 2-foot putt, and lost to Player by one stroke.

> LEON SPINKS HAS HIS HEAVYWEIGHT CHAMPIONSHIP TAKEN AWAY FROM HIM BY WORLD BOXING COUNCIL; CHARGED WITH NONCOOPERATION.
>
> | | Ph | Em | In |
> |---|---|---|---|
> | born: 7/11/53 | | | |
> | date: 3/18/78 | | | |
> | biorhythms: | ① | ① | 8 |

> F. LEE BAILEY'S BRILLIANT TRIAL WORK GOES FOR NAUGHT: THE "WOMAN SCORNED" GETS HER REVENGE.

F. Lee Bailey has a reputation for spectacular defenses in celebrated murder trials. He was able to get Ohio osteopath Dr. Sam Shepard acquitted 12 years after his conviction for murdering his wife. He cited prejudicial publicity as a basis for acquittal, because some jurors went home for the evening. There followed other well-covered trials, such as that of the Plymouth mail robbery suspects and the Boston Strangler, Albert DeSalvo.

On December 10, 1966, however, his legal gymnastics hit a paradoxical snag.

In Freehold, New Jersey, Dr. Carl Coppolino was on trial for murder by strangulation of Lt. Col. William E. Farber.

Marjorie Farber, wife of the victim, was a self-confessed lover of the alleged murderer.

BIOCYCLES 97

In a grandstand move to deflect attention away from the facts and, by implication, damage the testimony of the woman, Bailey handed a photo of Mrs. Farber to the jury. She is "the woman scorned," he cried. He was referring to what *could* have been her reaction to Coppolino's leaving her for another woman.

After the acquittal, it occurred to many that the doctor's wife had herself died, in August of 1975—forty days before Coppolino was to marry again.

In April of 1967, Dr. Coppolino was tried and convicted of murder. The jury determined that the physician injected succinylcholine chloride into his wife. It had been thought, until that trial, that the substance was undetectable in the human body.

F. Lee Bailey							
		Ph	Em	In	Ph	Em	In
born:	6/10/33	9	26	18	9	26	18
date:	12/10/66						
	4/15/67	15	3	32	3	17	3
biorhythms:		①	①	27	⑫	⑮	21

THE PHYSICAL/INTELLECTUAL COMBINATION

The problems involved in trying to perform logically at times when the body isn't cooperating bring about extraordinary states of confusion that might, someday, constitute legal defenses or excuses for inappropriate actions. It somewhat explains some of the more bizarre news events of the era.

ONE-TIME SUPREME COURT NOMINEE G. HARROLD CARSWELL ARRESTED ON HOMOSEXUALITY AND MORALS CHARGES

Tallahassee, FL, June 27, 1976—Judge G. Harrold Carswell, who was nominated for a seat on the U.S. Supreme Court by former President Richard Nixon, was arrested today. He is charged with attempting to commit a homosexual act with a vice squad officer. He pleaded guilty to the lesser charge of battery. Carswell had been nominated to the U.S. Supreme Court on January 19, 1970.

G. Harrold Carswell	Ph	Em	In
born: 12/22/19	6	17	20
date: 6/27/76	6	18	31
biorhythms:	⑫	7	⑱

As embarrassing as the Carswell story was, the publicity stirred up by the charge was not nearly as great as that caused by an event that took place two weeks later.

CARSWELL'S CLOSE FRIEND FOUND SLAIN

July 11, 1976—Jack Pack, a former high school teacher and friend of G. Harrold Carswell, was found slain. Pack was an avowed homosexual.

DR. J'S LAME: KNEE PROBLEM FLARES UP

New York, January 30, 1975—New York Nets basketball superstar pulled up lame from a running drill today, and complained of a sharp pain in his left knee. The flareup of tendonitis was his first since last year when he had to be fitted with a special brace. Recently he scored 21 points, 7 rebounds, and 7 assists in the ABA All-Star game victory of the East over the West.

Julius "Dr. J" Erving

	Ph	Em	In
born: 2/22/50	3	1	22
date: 1/30/75	21	8	12
biorhythms:	①	9	①

A look at Erving's setback is even more interesting when his biorhythms are checked for February 22, 1975, his very *next* physical critical day 1:

"Dr. J" SOARED ABOVE THE RIM; SCORES 51 POINTS AS NETS ROUT SAN DIEGO 126–93

San Diego, February 22, 1975—Clint Roswell of the *N.Y. Sunday News* summed it up well when he said Julius Erving "soared above the rim with as much fire and grace (that) after a while you had to wonder if Mother Nature would ever catch up with Dr. J."

	Ph	Em	In
born: 2/22/50	3	1	22
date: 2/22/75	21	3	2
biorhythms:	①	4	24

Other physical/intellectual combination days in biorhythm totals follow.

SEN. EDWARD M. KENNEDY LEAVES ACCIDENT SCENE; WOMAN PASSENGER DROWNS; KENNEDY SWIMS AND SUFFERS "MEMORY LAPSE."

	Ph	Em	In
born: 2/22/32			
date: 7/19/69			
biorhythms:	①	27	①

LEN DAWSON COMMITS MOST FUMBLES IN A SINGLE FOOTBALL GAME; DROPS 7 IN KANSAS CITY–SAN DIEGO GAME.

	Ph	Em	In
born: 6/20/35			
date: 11/15/64			
biorhythms:	①	18	⑰

BIOCYCLES   101

GOV. JERRY BROWN MAKES PASSIONATE "NO CAPITAL PUNISHMENT" ARGUMENTS; REBUKED BY FOES.	*Ph*	*Em*	*In*
born: 4/7/38			
date: 6/7/77			
biorhythms:	①	27	⑱

GERALD FORD SWORN IN AS PRESIDENT OF THE U.S.	*Ph*	*Em*	*In*
born: 7/14/13			
date: 7/9/74			
biorhythms:	⑫	16	①

ATTORNEY GENERAL GRIFFIN BELL FIRES CORRUPTION-FIGHTING U.S. ATTORNEY: SAYS HE'S DOING IT FOR "THE POLITICAL SYSTEM."

Washington, D.C., January 20, 1978

At the time of the David Marston firing in Philadelphia, the author ran several wire service dispatches over United Press International and the CBS Radio Network anticipating Bell's actions. At the time prior to January 20, 1978, there had been stories that U.S. Congressmen Joshua Eilberg (2/12/21) and Daniel Flood (11/26/04) had been involved in influence-peddling regarding the solicitation of funds for an addition to Hahnemann Hospital in that city.

Eilberg had telephoned Jimmy Carter in November 1977 and asked the President to "get rid of Marston."

David Marston and the FBI were investigating the "influence-peddling" stories at the time. Already there had been several political convictions in similar Philadelphia corruption cases.

When Marston went to Washington to "discuss his future" and status, the Bell reaction was biorhythmically predictable.

Griffin Bell

		Ph	Em	In
born:	10/31/18	9	14	8
date:	1/20/78	3	2	9
biorhythms:		(12)	16	(17)

SUPREME COURT JUSTICE WILLIAM H. REHNQUIST WRITES THAT PORNO ADVERTISING MAY BE USED IN OBSCENITY CASES; DENIES FIRST AMENDMENT PROTECTION.

		Ph	Em	In
born:	10/1/24			
date:	6/4/77			
biorhythms:		(12)	4	(1)

ELVIS PRESLEY DID *NOT* DIE ON A CRITICAL DAY.

Elvis Presley was conspicuously absent from the necrology in Chapter I: The reason is that he did *not* die on a critical day.

The story of the events *preceding* his death may help to explain this better.

On Sunday and Monday, less than two days before he died, he had *two* double critical days:

		Ph	Em	In
born:	1/8/35			
date:	8/14/77 Sunday			
	8/15/77 Monday			
biorhythms:		⑫	20	⑰
		⑬	21	⑱

There were few reports about the day immediately preceding his death except that he played tennis for 8 straight hours. His friends also indicated that he was on several potent drugs.

		Ph	Em	In
biorhythms:	8/16/77 Tuesday	14	22	19
	8/17/77 Wednesday	15	23	20

IV. The Odyssey of Triple Critical Days

They don't occur often but when they do the results can be major.

It definitely is the kind of day people *do* stay in bed. Cyrus Vance, for example, took it easy on his triple critical day in 1977. It was the only non-news day for him at a time when all hell was breaking loose in the SALT talks, a Pan American Union conference, and Middle East peace plans.

Vance spent his triple critical day watching a White House reception for visiting Israeli Prime Minister Begin.

Triple critical days occur when *all three* rhythms are either at the midpoint or at the beginning of their respective cycles. This happens only 13 times during an average lifetime.

The first is at birth, the second at 7.59 years, and the last, alas, at 65.77 years.

The 13 periods are illustrated on the following pages.

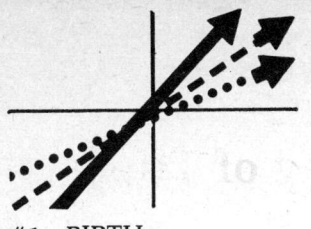

#1 BIRTH
0 year, 1 day
① – ① – ①

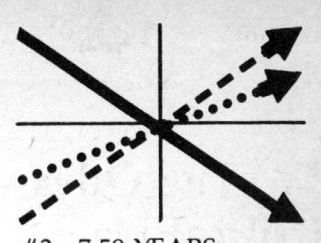

#2 7.59 YEARS
7 years, 215 days
⑬ – ① – ①
(2,773 days)

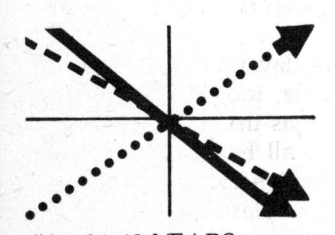

#3 21.49 YEARS
21 years, 185 days
⑫ – ⑮ – ①
(7,854 days)

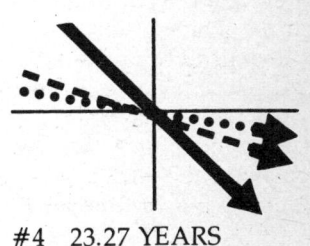

#4 23.27 YEARS
23 years, 97 days
⑫ – ⑮ – ⑱
(8,497 days)

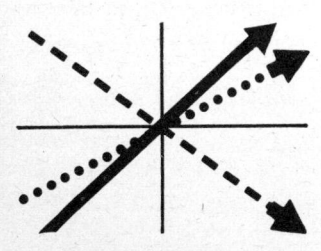

#5 29.09 YEARS
29 years, 35 days
① – ⑮ – ①
(10,626 days)

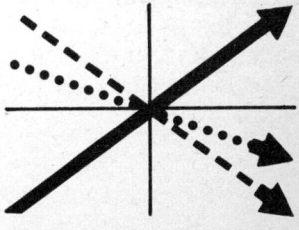

#6 30.85 YEARS
30 years, 314 days
① – ⑮ – ⑱

#7 34.93 YEARS
    34 years, 338 days
    ⑬ − ⑮ − ⑰

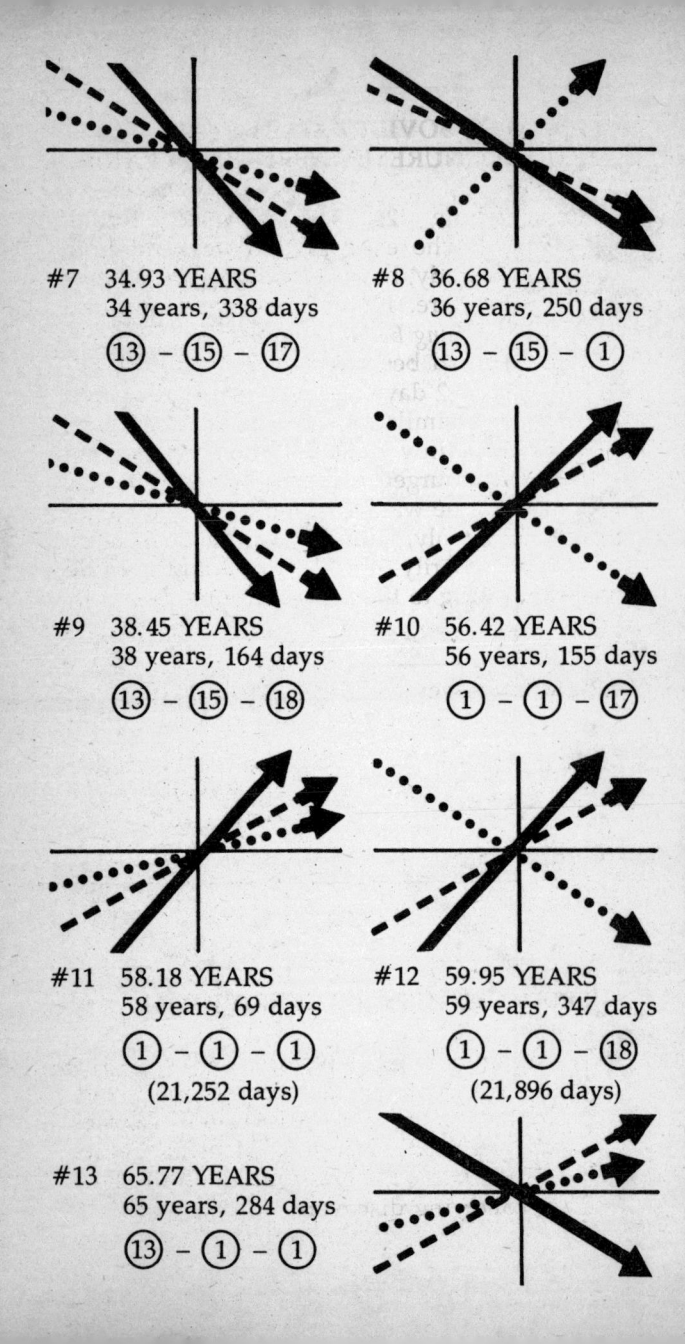

#8 36.68 YEARS
    36 years, 250 days
    ⑬ − ⑮ − ①

#9 38.45 YEARS
    38 years, 164 days
    ⑬ − ⑮ − ⑱

#10 56.42 YEARS
    56 years, 155 days
    ① − ① − ⑰

#11 58.18 YEARS
    58 years, 69 days
    ① − ① − ①
    (21,252 days)

#12 59.95 YEARS
    59 years, 347 days
    ① − ① − ⑱
    (21,896 days)

#13 65.77 YEARS
    65 years, 284 days
    ⑬ − ① − ①

ESCAPED SOVIET BALLET STAR RUDOLF NUREYEV APPEARS IN PARIS

Paris, June 22, 1961—Dancer Rudolf Nureyev, who escaped this weekend from Soviet custody, has been retained by a ballet company here. He will perform Florimund in *The Sleeping Beauty* tomorrow.

Nureyev had been told to return to the Soviet Union 2 days ago for "insubordination and non-assimilation." He had been touring with the Kirov dancers when he escaped from Le Bourget Airport. At the time the dance troupe was boarding a plane for London. Suddenly, Nureyev was singled out by Soviet "security officials" and told he would be returning to the Soviet Union.

Rudolf Nureyev

		Ph	Em	In
born:	3/17/38	6	21	26
date:	6/22/61	6	22	25
biorhythms:		⑫	⑮	⑱

FIRST LADY IS A CANCER VICTIM

Washington, D.C., September 28, 1974—The White House today announced that Betty Ford has a malignant breast tumor. Doctors at the Bethesda National Medical Center performed a radical mastectomy on the First Lady following discovery.

Mrs. Ford revealed later that the month of September 1974 was a "painful period in our lives." Mrs. Ford suffered a triple critical day on September 9, 1974.

		Ph	Em	In
born:	4/8/18	8	24	16
date:	9/9/74	16	5	1
biorhythms:		①	①	⑰

PLAYER CHARGED IN ASSAULT

Cincinnati, Oct. 1977—Reds pitcher Pedro Borbon was arrested and is awaiting arraignment on a felony assault charge at a discotheque . . .

Pedro Borbon

		Ph	Em	In
born:	12/2/46	8	3	12
date:	10/10/77	16	12	6
biorhythms:		①	⑮	⑱

BIOCYCLES 109

SCIENTIST HOWARD M. GOODMAN USES RECOMBINANT DNA TO REPRODUCE THE GENE INSULIN.

Howard M. Goodman			
	Ph	Em	In
born: 11/29/38	12	16	0
date: 5/10/77	1	27	18
biorhythms:	⑬	⑮	⑱

(Reported 5/23/77 in all wire services.)

UPCOMING TRIPLE CRITICAL DAYS

Using the time information that determines when a triple critical day is upcoming, any person approaching the given number of days alive may be predicted for such a day.

The month for the triple critical day may be readily determined by finding the series of bio-values on the date tables that best corresponds to those of the given person's.

An example:

Stevie Wonder

	Ph	Em	In
born: 5/13/50	15	5	8
nearest date's bio-values:			
6/--/79	17	23	11
add 15 days	15	15	15
6/16/79 -6*			
	(23)	(28)	(33)
	1	15	1

In the above situation, number differences are sought as follows:
- Ph — 2 days greater
- Em —18 days greater
- In — 3 days greater

---

*The numerical notation indicates the day of the week. This is found by consulting a long-range or perpetual calendar, or by using the *Kosmos* biorhythm calculator.

TRIPLE CRITICAL DAYS

1979 Triple Critical Days

Date		Name	Ph	Em	In
9/19	-3	Pierre Trudeau (10/18/19)	1	1	18
10/28	-7	Henry Kissinger (5/27/23)	1	1	17

1980

8/28	-4	Bill Blass (6/22/22)	1	1	1
9/24	-3	George McGovern (7/19/22)	1	1	1

1981

11/13	-5	Peter Lawford (9/7/23)	1	1	1

1982

6/30	-3	George McGovern (7/19/22)	1	1	18
12/8	-3	Jimmy Carter (10/1/24)	1	1	1

1983

2/26	-6	Marjie Wallace (1/23/54)	1	15	1
4/17	-7	Jack Lemmon (2/8/25)	1	1	1
8/19	-5	Peter Lawford (9/7/23)	1	1	18

1984

1/22	-7	Howard Baker (11/15/25)	1	1	1
1/24	-2	Chris Evert (12/21/54)	1	15	1

1985

6/3	-1	George Wallace (8/25/19)	13	1	1
7/6	-6	Charles Percy (9/27/19)	13	1	1
7/17	-3	Omar Torrijos (2/13/29)	1	1	17
8/4	-7	Edward Brooke (10/26/19)	13	1	1
11/9	-6	Paul Warnke (1/31/20)	1	1	1

1986

2/26	-3	Princess Caroline (1/23/57)	1	15	1
3/13	-4	Walter Mondale (1/5/28)	1	1	1
4/15	-2	Princess Grace (11/12/29)	1	1	17
5/3	-6	Bella Abzug (7/24/20)	13	1	1

V. The Biorhythm Ping-Pong of Diplomacy

THE SALT TALKS

When Jimmy Carter assumed the U.S. Presidency in 1977, most observers appreciated his inexperience in foreign affairs and even speculated about the dangers of stalemate in important international negotiations.

During the first months of his administration, the major agenda item was the reconvening of the Strategic Arms Limitation (SALT) talks between the Soviet Union and the United States.

BIOCYCLES 113

> **SALT TALKS BREAK DOWN IN MOSCOW**
>
> Washington, Mar. 25, 1977—President Carter, in a hastily arranged news conference, after conferring with Congressional leaders on breakdown of Secretary Vance's arms limitation talks in Moscow, says he's not discouraged . . .
> —from the *New York Times*

The principal participants in this news story were:

    Jimmy Carter and Leonid Brezhnev
    Born 10/1/24   Born 12/19/06

Both agreed to convene a meeting in Moscow in late March 1977. Brezhnev and Andrei Gromyko (born 7/6/09) would meet with Carter's designates. The problem was that Carter's Secretary of State, Cyrus Vance, had only recently been confirmed by the Congress. Also, Carter's negotiator, Paul Warnke, was not confirmed until two weeks before the meeting!

The reasons for these defects in the negotiating plan, and for the consequent events during March, become evident when the biorhythms of the participants are paired with their actions and inactions. The dates and statements selected represent the total coverage of the month's events in major U.S. newspapers.

A biorhythm mapping of the news development dramatically shows *why* the talks could probably not have succeeded, and *why* no one was apparently in a position to realize they wouldn't. Note the critical days (circled).

Date	Jimmy Carter	Cyrus Vance	Paul Warnke	Andrei Gromyko	Leonid Brezhnev
—	b. 10/1/24	b. 3/27/17	b. 1/31/20	b. 7/6/09	b. 12/19/06
	bv. 9-8-24	bv. 17-9-30	bv. 11-4-12	bv. 9-3-13	bv. 19-8-19
3/8/77		Decides to bring his own interpreter instead of using the Soviets' as had Kissinger. ① – ① – ⑱			
3/14/77	Begins human rights campaign despite advice that this will jeopardize the negotiations. 22 – 6 – ⑱				

Date	Jimmy Carter	Cyrus Vance	Paul Warnke	Andrei Gromyko	Leonid Brezhnev
	b. 10/1/24	b. 3/27/17	b. 1/31/20	b. 7/6/09	b. 12/19/06
—	bv. 9-8-24	bv. 17-9-30	bv. 11-4-12	bv. 9-3-13	bv. 19-8-19
3/25/77			Newly confirmed negotiator does not expect that rights issue will affect outcome of arms talks. 13 – 14 – ⑱		
3/28/77				Bans 30 U.S. newspeople from 1st day of talks because room would be too "crowded." ⑬ – 14 – 21	

Date	Jimmy Carter	Cyrus Vance	Paul Warnke	Andrei Gromyko	Leonid Brezhnev
—	b. 10/1/24 bv. 9-8-24	b. 3/27/17 bv. 17-9-30	b. 1/31/20 bv. 11-4-12	b. 7/6/09 bv. 9-3-13	b. 12/19/06 bv. 19-8-19
3/29/77					Surprises the world by not showing up for the talks. ① – 21 – 28
3/30/77	Confirms "breakdown" of SALT talks. Says he's not discouraged. 15 – 22 – ①				

DEVELOPING STORY: Salt Negotiations
PARTICIPANTS: Carter, Vance, Warnke, Gromyko, Brezhnev
BIRTH DATA: 10/1/24 3/27/17 1/31/20 7/6/09 12/19/06

DATES:	P E I	P E I	P E I	P E I	P E I
3/8/77		1 1 18			
3/14/77	22 6 18				
3/25/77			13 14 18		
3/28/77				13 14 21	
3/29/77					7 21 26
3/30/77	15 22 1				

PEOPLE IN GOVERNMENT SERVICE—U.S.

		Ph	*Em*	*In*
Gardner Ackley	6/30/15	9	1	6
Brock Adams	1/13/27	3	14	15
Carl Albert	5/10/08	17	4	6
Clifford L. Alexander, Jr.	9/21/33	21	7	14
Cecil Andrus	8/25/31	20	9	13
Howard Baker	11/15/25	13	18	10
Griffin Bell	10/31/18	9	14	8
Bob Bergland	7/22/28	22	18	20
W. Michael Blumenthal	1/3/26	11	25	27
Julian Bond	1/14/40	14	24	17
Thomas Bradley	10/29/17	8	17	12
Zbigniew Brzezinski	3/28/28	0	22	4
William E. Brock	11/23/30	19	4	24
Edward W. Brooke	10/26/19	17	18	11
George Brown	8/17/18	15	5	17
Harold Brown	9/19/27	7	17	30
Jerry Brown, Jr.	4/7/38	18	28	5
Warren Burger	4/21/05	5	27	32
Joseph Califano, Jr.	5/15/31	7	27	16
Hugh Carey	4/11/19	21	20	11
Ramsey Clark	12/18/27	9	11	6
Max Cleland	8/24/42	5	24	22
John B. Connally	2/28/17	21	8	24
Midge Costanza	11/28/32	10	24	14
Carl T. Curtis	3/15/05	19	8	3
Robert J. Dole	9/22/23	9	25	32
Thomas Eagleton	9/4/29	4	1	7
James Eastland	11/28/04	11	3	11
Robert Finch	10/9/25	5	28	15
Gerald Ford	7/14/13	12	17	29
William Fulbright	4/9/05	17	11	11
Barry Goldwater	1/1/09	6	20	1
Robert Griffin	11/6/23	18	3	25
Alexander Haig	12/2/24	15	2	28
Fred R. Harris	11/13/30	4	14	1
S. I. Hayakawa	7/18/06	12	22	8

PEOPLE IN GOVERNMENT SERVICE—U.S.

		Ph	Em	In
Harold Hughes	2/10/22	7	20	31
Hubert H. Humphrey	5/27/11	9	12	16
Henry M. Jackson	5/31/12	7	6	9
Jacob Javits	5/18/04	21	1	7
Leon Jaworski	9/19/05	15	16	13
Barbara Jordan	2/21/36	11	19	21
Hamilton Jordan	9/21/44	5	21	22
Irving Kaufman	6/24/10	1	13	21
Jack F. Kemp	7/13/35	5	19	14
Edward M. Kennedy	2/22/32	22	23	29
Henry Kissinger	5/27/23	19	25	22
Edward I. Koch	12/12/24	6	20	31
Juanita M. Kreps	1/11/21	11	23	30
Bert Lance	6/3/31	11	8	30
Russell B. Long	11/3/18	6	11	5
Clare Boothe Luce	4/10/03	11	13	15
Richard Lugar	4/4/32	4	10	21
Michael Mansfield	3/16/03	13	10	7
Freddie Marshall	8/22/28	14	15	22
Ray Marshall	8/22/28	14	15	22
Paul N. McCloskey	9/29/27	20	7	20
James A. McClure	12/27/24	14	5	3
George S. McGovern	7/19/22	9	1	4
Thomas J. McIntyre	11/4/07	21	24	29
Wm. G. Milliken	3/26/22	9	16	1
John Mitchell	9/15/13	18	10	32
Walter F. Mondale	1/5/28	13	20	20
Daniel Patrick Moynihan	3/16/27	10	8	19
George C. Murphy	7/4/02	15	13	31
John M. Murphy	8/3/26	5	9	13
Edmund Muskie	3/28/14	8	12	3
Gaylord Nelson	6/4/16	14	25	29
Richard M. Nixon	1/9/13	14	7	17
Charles H. Percy	9/27/19	23	19	7
Carl Perkins	10/15/12	8	9	4
Jody Powell	9/30/43	17	14	16

PEOPLE IN GOVERNMENT SERVICE—U.S.

		Ph	Em	In
Lewis Powell	9/19/07	21	16	9
Thomas Railsback	1/22/32	7	26	24
William Randall	2/16/09	11	2	21
Jennings Randolph	3/8/02	18	19	17
Charles Rangel	6/11/30	23	1	24
Robert D. Ray	9/26/28	2	8	20
Ronald Reagan	2/6/11	4	10	27
Thomas Rees	3/26/25	17	28	13
John J. Rhodes	9/18/16	23	3	22
Elliot Richardson	7/20/21	5	1	5
Nelson Rockefeller	7/8/08	4	11	13
James Schlesinger	2/15/29	21	6	10
Richard Schweiker	6/1/26	22	16	10
Hugh Scott	11/11/00	17	25	4
Milton Shapp	6/25/12	5	9	17
Tom Stafford	9/17/30	17	5	25
James Symington	9/28/27	21	8	21
Cyrus R. Vance	3/27/17	17	9	30
George Wallace	8/25/19	10	24	7
Paul Warnke	1/31/20	11	4	12
Walter Washington	4/15/15	16	21	16
Richard Wiley	7/20/34	18	13	9
Andrew Young	3/12/32	4	5	11
Coleman Young	5/24/18	8	24	3

INTERNATIONAL NEWSMAKERS

		Ph	Em	In
Hamilton Shirley Amerasinghe	3/18/13	15	23	15
Raymond Barre	4/12/24	20	12	31
Leonid Brezhnev	12/19/06	19	8	19
James Callaghan	3/27/12	3	15	8
Fidel Castro	8/13/27	21	26	1
Raúl Castro	6/3/31	11	8	30
Moshe Dayan	5/20/15	4	14	14
Brian Faulkner	2/18/21	19	13	25
Indira Gandhi	11/19/17	10	24	24
V. Giscard d'Estaing	2/2/26	3	23	30
Emperor Hirohito	4/29/01	9	24	33
Juan Carlos I	1/5/38	18	8	31
Helmut Kohl	4/3/30	23	14	27
René Lévesque	8/24/22	19	21	1
Gaafar M. al-Nimeiry	1/1/30	23	22	20
David Owen	7/2/38	1	26	18
Mohammad Reza Shah Pahlavi	10/26/19	1	20	8
Chung Hee Park	11/14/17	15	1	29
José López Portillo	6/16/20	13	8	8
Anwar Sadat	12/25/18	23	15	19
Anastasio Somoza Debayle	12/5/25	16	26	23
Adolfo May Suárez González	9/25/32	14	4	12
Nguyen Van Thieu	4/5/23	2	21	8
Leo Tindemans	4/16/22	11	11	32
Omar Torrijos Herrera	2/13/29	0	8	12
Pierre Trudeau	10/18/19	2	26	19
Harold Wilson	3/11/16	7	26	15

OTHER NEWSMAKERS

		Ph	Em	In
Ralph Abernathy	3/11/26	12	14	26
Bella Abzug	7/24/20	21	26	3
Spiro Agnew	11/9/18	23	5	32
Saul Alinsky	1/30/09	5	19	5
George Allen	4/29/22	21	26	19
Sparky Anderson	2/22/34	5	21	25
Isaac Asimov	1/2/20	12	5	9
Richard Avedon	5/15/23	8	9	1
Richard Bach	6/23/36	4	9	31
George Baldwin	5/5/17	1	21	24
James Beard	5/5/03	9	16	23
Saul Bellow	6/10/15	6	21	26
Peter Benchley	5/8/40	15	22	2
Elizabeth Bishop	2/8/11	2	6	25
Jim Bishop	11/21/07	4	7	12
Baruch Blumberg	7/28/25	8	16	21
Erma Louise Bombeck	2/21/27	10	3	9
Daniel J. Boorstin	10/1/14	5	21	14
Jimmy Breslin	10/17/30	10	13	28
Gwendolyn Brooks	6/7/17	19	21	24
Susan Brownmiller	2/15/35	15	27	30
William Buckley	10/6/13	20	17	11
Cab Calloway	12/24/07	17	2	12
Truman Capote	9/30/24	10	9	25
Cesar Chavez	3/31/27	18	21	4
John Cheever	5/27/12	11	10	13
Charles Colson	10/16/31	14	13	27
Norman Cousins	6/24/12	6	10	18
Jacques Cousteau	6/11/10	14	26	03
Michael De Bakey	9/7/08	12	24	18
John Ehrlichman	3/20/25	0	6	19
Ralph Ellison	3/1/14	12	11	30
Werner Erhard	9/5/35	20	21	26
Princess Grace	11/12/29	4	16	4
Billy Graham	11/7/18	2	7	1
H. R. Haldeman	10/27/26	12	8	27
Alex Haley	8/11/21	6	7	16
William Hanna	7/14/10	4	21	3

OTHER NEWSMAKERS

		Ph	Em	In
Huntington Hartford	4/18/11	2	23	22
Richard Helms	3/30/13	3	11	3
Jane Jacobs	5/1/16	2	3	30
Robert Jastrow	9/7/25	13	3	13
Bruce Jenner	10/28/49	5	6	7
Lady Bird Johnson	12/22/12	9	25	2
Coretta Scott King	4/27/27	14	22	10
Evel Knievel	10/17/38	9	3	10
Mark Lane	2/24/27	7	0	6
Norman Lear	7/27/22	1	21	4
Claudine Longet	1/29/42	5	7	31
James Martin	6/21/20	8	3	3
Floyd McKissick	3/9/22	3	21	4
Russell Means	11/10/39	11	6	17
James Michener	2/3/07	19	18	6
Ralph Nader	2/27/34	0	16	20
Joe Namath	5/31/43	1	24	6
Leroy Neiman	6/8/26	15	9	3
Rudolf Nureyev	3/17/38	16	21	26
Madalyn Murray O'Hair	4/13/19	6	18	9
Anthony Powell	12/21/05	14	7	19
Howard Richardson	12/2/17	20	11	11
Oral Roberts	1/24/18	13	14	24
Pete Rozelle	3/1/26	22	24	3
Jonas Salk	10/28/14	8	24	17
William Saroyan	8/31/08	19	3	25
Alan Shepard	11/18/23	5	18	12
Jean Shepherd	7/26/29	21	13	14
B. F. Skinner	3/20/04	11	4	33
Aleksandr Solzhenitsyn	12/11/18	14	1	33
Stephen Spender	2/28/09	22	18	9
William Styron	6/11/25	9	7	2
Kurt Vonnegut	11/11/22	9	26	21
Jessamyn West	7/18/02	24	27	17
Stevie Wonder	5/13/50	15	5	8

WHO WILL BE THE NEW NEWSMAKERS OF THE 80'S?

Just as it's very difficult to remember Tiny Tim, Twiggy, and others of the past decade, it's doubly hard to anticipate who the *new* newsmakers of the future are likely to be.

Here are some up-and-coming people who provide a take-off point. There are spaces between the names to fill in additional names as new fads and faces emerge.

		Ph	Em	In
Christina Onassis Andreadis	12/11/50	10	17	27
Elizabeth Ashley	8/30/39	14	22	33
Judith Blegen	4/27/41	6	4	11
Julian Bond	1/14/40	14	24	17
Vladimir Bukovsky	12/30/42	15	8	26
Ed Bullins	7/2/35	16	2	25
Rod Carew	10/1/45	0	10	10
Princess Caroline	1/23/57	6	22	3
Steve Cauthen	5/1/60	8	4	30
Nadia Comaneci	11/12/61	0	4	31
Larry Csonka	12/25/46	8	8	22
Mark Fidrych	8/14/54	2	19	5
Douglas Fraser	12/18/16	1	24	30
Mirella Freni	2/27/35	3	15	18

Felipe Gonzalez	3/5/42	16	0	29
Reggie Jackson	5/18/46	22	5	12
Marcel Lefebvre	11/29/05	13	1	8
Delano Meriwether	4/23/43	16	6	11
Bette Midler	12/1/45	6	5	15
Lee Bouvier Radziwill	3/3/33	16	13	18
John D. Rockefeller IV	6/18/37	12	13	1
Marjie Wallace	1/23/54	21	26	10
André Watts	6/20/46	12	0	12

Other Readings on Biorhythms and Performance Correlates

Ahlgren, Andrew, "Labor Investigation of 'Biorhythms'" (a letter), *Aviation Space and Environmental Medicine*, 48(6):678, July, 1977.

Barrows, Charles H., Jr., et al., "Protein Synthesis, Development, Growth, and Life Span," *Growth*, 39(4):525–533, December, 1975.

Botwinick, J., et al., "Qualitative Vocabulary Test Responses and Age," *Journal of Gerontology*, 30(5):574–577, September, 1975.

Bouma, J. J., and Tromp, S. W., "Daily, Monthly, and Yearly Fluctuations in Total Number of Suicides and Suicide Attempts in the Western Part of The Netherlands," *Journal of Interdisciplinary Cycle Research*, 3(3–4), 1972.

Brown, Frank A., Jr., "Biological Clocks: Endogenous Cycles Synchronized by Subtle Geophysical Rhythms," *Biosystems*, 8(2):67–81, July, 1976.

Caille, E. J., et al., "Biorhythms and Watch Rhythms: Hemeral Watch Rhythms and Anhemeral Watch Rhythms in Simulated Permanent Duty," in Mackie, R. R., ed., *Vigilance: Theory, Operational Performance, and Physiological Correlates*, New York, Plenum, 1977.

Delgado, J., *Proceedings of the 23rd International Congress of Physiological Sciences*, Tokyo, 1965.

Deyo, M. E., "Is Today for the Birds or the Birdies?" *The Professional Golfer*, April, 1976.

Doering, C. G., et al., "A Cycle of Plasma Testosterone in the Human Male," *Journal of Clinical Endocrinological Metabolism*, 40(3):492–500, March, 1975.

Doskin, V. A., et al., "Biorhythmical Prerequisites for Organizing the Shiftwork of Young Workers," *Gigiena trudi i professionalnye zabolevaniia*, 18(4):9–13, April, 1976.

Doskin, V. A., et al., "Physiological Rhythms and the Work Capacity of Man," *Gigiena i sanitariia* (38):76–91, August, 1973.

Dreiske, Paul, "Strange Forces in Our Times," *Family Safety*, 15, Summer, 1972.

Gittelson, Bernard, *Biorhythm Sports Forecasting*, New York, Arco Publishing, 1977.

Glasser, Stanley R., "Biological Rhythmicity Influencing Hormonally-Inducible Events," *Methods in Enzymology*, 36(00):474–481, 1975.

Gross, Hugo Max, *Biorhythmik: das auf und ab unseres Lebens, Einfuhrung und Anleitung*, Freiburg, Hermann Bauer Verlag, 1959.

Kripke, Daniel F., and Lavie, Peretz, "Ultradian Rhythms: The 90-Minute Clock Inside Us," *Psychology Today*, 65:55–56, April, 1975.

Leaf, A., "Every Day Is a Gift When You Are over 100," *National Geographic*, 143(1):118, 1973.

Lipa, B. J., et al., "Search for Correlation Between Geomagnetic Disturbances and Mortality," *Nature* (259):302, 1976.

Lovett-Doust, J. W., et al., "Comparison Between Some Biological Clocks Regulating Sensory and Psychomotor Aspects of Perception in Man," *Neuropsychobiology*, 1(5):261–266, 1975.

Mallardi, Vincent, *Biorhythms & Your Behavior*, rev. ed., Philadelphia, Running Press, 1978.

Menaker, Michael, ed., *Biochronometry*, Washington, D.C., National Academy of Sciences, 1971.

Mills, J. N., ed., *Biological Aspects of Circadian Rhythms*, London, Plenum Press, 1973.

O'Neil, Barbara, and Phillips, Richard, *Biorhythms: How to Live with Your Life Cycles*, Pasadena, Ward Ritchie Press, 1975.

Orr, W., Letter: REM Sleep and Cardiac Arrhythmias, *Circulation*, 52(3):519, September, 1975.

Pessacq, M. T., et al., "Effect of Fasting on the Circadian Rhythm of Serum Insulin Levels," *Chronobiologia*, 2(3):205–209, July–September, 1975.

Podnieks, I., et al., "Characteristics of a Neural Clock Regulating Perception and Psychomotor Performance in Man," *Biological Psychology*, 4(4):265–276, December, 1976.

Randel, Hugh W., ed., *Aerospace Medicine*, Baltimore, Williams and Wilkins, 1970.

Reimann, Hobart A., "Rhythms and Periodicity in Health and Disease," *Annals of Clinical Laboratory Science*, 5(6):417–420, 1975.

Reinberg, Alain, et al., "Circannual and Circadian Rhythms in Plasma Testosterone in Five Healthy Young Parisian Males," *Acta Endocrinologia*, 80(4):732–734, December, 1975.

Richter, Curt Paul, *Biological Clocks in Medicine and Psychiatry*, Springfield, IL, Thomas, 1965.

Schara, August, *All the Presidents Plus: An Insight into Moments of History Through Biorhythms*, Hicksville, NY, Exposition Press, 1978.